HOW NUCLEAR WEAPONS DECISIONS ARE MADE

HOW NUCLEAR WEAPONS DECISIONS ARE MADE

Edited by Scilla McLean

John Beyer, Julian Cooper,
Gerard Holden, François Nectoux,
Nancy Ramsey, David Schorr,
Tony Thomson, Andrew White

M
MACMILLAN in association with THE OXFORD RESEARCH GROUP

First published 1986

Published by
THE MACMILLAN PRESS LTD
Houndmills, Basingstoke, Hampshire RG21 2XS
and London
Companies and representatives
throughout the world

Typeset by Latimer Trend & Company Ltd, Plymouth

Printed in Great Britain by
Anchor Brendon Ltd, Tiptree, Essex

British Library Cataloguing in Publication data

How nuclear weapons decisions are made
1. Nuclear weapons—Government policy
I. McLean, Scilla ii. Oxford Research Group
355′.0335 UA11
ISBN 0–333–40582–X (hardcover)
ISBN 0–333–40583–8 (paperback)

Contents

List of Tables

List of Figures

Preface

The studies on which this book is based were first undertaken in 1982 by non-specialists, ordinary citizens who began to ask themselves some of the questions which appear on the first page. The work carried out at that stage was done by Danielle Artman, Nancy Ramsey, David Schorr and Tony Thomson. They were later joined by specialists: the chapter on the Soviet Union was written by Julian Cooper, the chapter on China by John Beyer, the chapter on NATO by Andrew White and the chapter on the Warsaw Treaty Organisation by Gerard Holden. The chapter on the United States was written by Scilla McLean drawing on the work of Nancy Ramsey and Mark Rovner, the French chapter by François Nectoux drawing on the work of David Schorr and the chapter on Britain by Scilla McLean drawing on information gathered by Tony Thomson and MacDonald Graham.

In preparing this book we have received the generous help of many scholars. We are grateful to David Holloway who read and commented on the Soviet and Warsaw Treaty Organisation chapters, to Milton Hoenig who scrutinised the US chapter for factual errors, to Graham Spinardi who contributed material for the case-study on Chevaline, to Agnes Bertrand and Jean Pharabod who documented specified topics in the French chapter, to Malcolm Spaven who commented on the NATO chapter, to Paul Rogers, Malcolm Dando and Hugh Miall who read and criticised the entire manuscript and to John Hamwee who made useful structural suggestions.

We have also received assistance from many of the organisations described in the book including, among others, the Directorate for Organisational and Management Planning in the Office of the Secretary of Defence and the Department of Energy Weapons Programme in Washington, the Office of the Permanent Under Secretary at the Ministry of Defence in Whitehall, the Defence Policy and Materiel Division at the Treasury, and the Arms Control and Disarmament Department at the Foreign and Commonwealth Office, the staff of the House of Commons Defence Select Committee and the House of

Commons Library, the Service d'Information et des Relations Publiques des Armées and the Ministère de Défense in Paris, and the Public Information Branch of Supreme Headquarters Allied Powers Europe. Some of the bodies concerned have been open and forthcoming with accurate information, others less so. Naturally, none of those who have been kind enough to assist us in our task bears any responsibility for the overall picture we have put together – that is entirely our own. But without the expert help of those concerned we would not have been able to draw together a first broad outline of how nuclear weapons decisions are made in five nations. Like any early map or chart, it will inevitably need refining, and we welcome comments and criticism which will improve future editions.

We would like to thank Julie Hudson who has patiently managed the word processing of successive drafts with great good humour, Rosie Houldsworth, Margot Miller and Janet Dando for help with proof-reading, and Anne Piper for providing research assistance.

We are very appreciative of the financial support provided by the Joseph Rowntree Charitable Trust, the Barrow and Geraldine S. Cadbury Trust, the Resource Group and the Noel Buxton Trust which enabled us to undertake the research.

Finally, we thank our families for their support and their good questions without which the book would not have come about. We dedicate it to them and to the families of all those upon whose decisions our future rests.

Woodstock, July 1985

Notes on the Contributors

John Beyer, a Cambridge graduate, is now presenting his Ph.D. thesis at Leeds University on Chinese peasant writers; he was formerly Asia Researcher at Amnesty International, after a year at Peking Languages Institute.

Julian Cooper, Lecturer in Soviet Technology and Industry, Centre for Russian and East European Studies, University of Birmingham. His published works include studies of Soviet science policy, technology and industry, including the defence sector.

Gerard Holden, a graduate of University College, Oxford and of Sussex University. He is currently Research Officer at the Science Policy Research Unit, Sussex University.

François Nectoux, holds a Ph.D. in economics and political science from the Sorbonne and is a graduate of the Institute of Political Studies in Paris. He is currently an Economist and Resource Analyst for Earth Resources Research.

Nancy Ramsey, gained her M.A. in politics from the University of Chicago, co-founded the Committee for National Security with Paul Warnke, and is currently Policy Director for Senator John Kerry.

David Schorr, a graduate of Oberlin College, has completed his first year of postgraduate law studies at Yale University.

Tony Thomson, who designed the original diagrams and flow charts which illustrate the text, trained at the City and Guilds of London Art School.

Andrew White, is a postgraduate student in the Department of International Relations at the London School of Economics, where his specialised subject is the Strategic Studies Community.

Introduction

There are three reasons why this book has been written.

Public concern over nuclear weapons has focused on the deployment of nuclear weapons, West and East. But deployment is merely the ultimate stage in a process which may have taken twenty years. Decisions made in the late 1960s have determined the kind of nuclear weapons we have in our arsenals today. Decisions made today determine the kind of weapons we will have at the turn of the century. Upon the outcome of these decisions, upon whether they are good decisions or bad, wise or foolish, the future of the earth now depends. Yet up to now these decisions have been shrouded in obscurity, and very little is known about how they are taken. On what information are they based? By whom are they taken? What kind of advice is sought? Which are the crucial decisions? If a decision is a bad one, how is it un-made? To answer some of these questions, it is necessary to unravel the complex processes of decision-making; to understand these processes, it is necessary to describe the structure – the organisations, institutions, councils and committees which control the process. This book is intended as a first step in that understanding.

Secondly, if citizens do not know how decisions on nuclear weapons are arrived at, they cannot be satisfied that the process is properly accountable. The degree of secrecy which has surrounded the development of nuclear weapons for forty-five years is one of the main causes of insecurity and alarm over current deployments. While a certain level of classification on key defence issues is clearly necessary, a great deal more information could be made available to the public with the result, desirable to all, that the machinery of democracy could take its proper course.

The third intent of the book is to provide a balanced account of nuclear decision-making, West and East, and thus a basis for future comparison and analysis of what happens in the different nations. A one-sided view of Western decision-making is of little interest, and for this reason equal attention has been paid to how the process works in the Soviet Union and China.

One brief general observation is in order at the outset of the book. It concerns the difference between nuclear weapons policy-making, and decisions on individual weapons systems. The nuclear weapons policy of a particular nation may remain constant over long periods of time. Britain's nuclear policy, for example, has remained essentially unchanged since the early 1960s, through four changes of party in government. But policy tends to become augmented, is certainly perpetuated, and may even be dictated, by decisions on indvidiual weapons systems over time. The paradox is that policy options and policy alternatives are hardly ever under scrutiny when decisions on individual weapons systems are being shaped. By the time these decisions have effectively been taken, it is often too late or too difficult to give serious consideration to policy change. It may be useful to bear in mind this distinction when using the book.

This book is not intended as a scholarly analysis of decision-making, but rather as a manual or reference book for those who want to understand the basic details of decision-making on nuclear weapons, not only in Britain but in the other nuclear nations and the major alliances. It describes the principal government departments which play a part in the process, the scientific institutions, military commands, manufacturing industries, advisory councils, cabinets and committees. It describes the stages through which a weapons system must pass on the way to deployment in each of the countries, and where appropriate illustrates this process with a case study of a particular weapon. It is not a technical guide to weapons development, nor does it offer any historical account of decision-making – it is concerned largely with the situation in the 1980s. The arms control negotiating process is not discussed in any detail, nor is the influence of the media, or of political parties.

The book opens with chapters on nuclear weapons decision-making in each of the superpowers. The next chapter concerns Britain and contains, for the British reader, more detail on organisational and departmental structures than will be found in other chapters; the net had deliberately been cast wide, so that the reader is aware of the full range of influences. In later books a more analytical assessment of the situation in Britain may be possible, as the picture becomes clearer. The chapter on France concentrates on the very early stages of weapons development when the crucial decisions are made. In the case of China, the author confined himself to the limited factual information available, and resisted any tendency towards speculation. The chapter on NATO brings together some of the points from earlier chapters on the

Western nations, illustrating how membership of an alliance can influence national decision-making. The final chapter on the Warsaw Treaty Organisation depicts a different kind of alliance and the interactions between member nations.

The authors have relied exclusively on unclassified sources and upon published information. Because of the extent of secrecy on these issues, there will obviously be gaps and omissions. But the book is a first, and for the first time the public has a map of the obscure and hitherto uncharted world of nuclear decision-making.

1 The Soviet Union

In August 1949 the Soviet Union tested its first atomic bomb, four years later, a thermonuclear device, and in 1957 the world's first intercontinental ballistic missile. In 1956 the USSR began deploying strategic bombers capable of reaching the United States. Thus during the first post-war decade the Soviet leadership mobilised the scientific and technical forces of a country grievously scarred by war in an effort to break the Western nuclear monopoly. The Soviet nuclear and missile programmes were undertaken with highly centralised forms of management with close involvement of the top Communist Party and government leadership in conditions of the strictest secrecy. Out of this experience emerged a set of institutions, and beliefs, which thereafter underwent only gradual evolution. It was not until the 1970s that a position of broad strategic parity with the USA was attained and there began a process of reassessment of the military role of nuclear weaponry. The top political and military leadership began to stress publicly the unwinnability of nuclear war and disavow the quest for nuclear superiority. According to the new Constitution adopted in 1977, 'The state ensures the security and defence capability of the country, and supplies the armed forces of the USSR with everything necessary for that purpose'.[1] With respect to nuclear weapons, the 'everything necessary' appears now to be assessed in terms of the maintenance of a state of parity with the West within a framework of mutual deterrence.[2]

The 1977 Constitution also appears to have regularised the formal status of some of the decision-making bodies concerned with national security. In this chapter an attempt is made to review this institutional structure and identify the principal forces which shape Soviet nuclear policy. The main bodies with which we are concerned are the Communist Party, the state and government, the military, and the defence industry with its research and design organisations responsible for the development of weapons systems. Figure 1.1 shows a schema of the national security decision-making structure. It must be emphasised that decision-making for defence in the USSR is shrouded in secrecy, and nowhere more so than in relation to nuclear weapons. Virtually nothing is known about the research and production organisations of the nuclear weapons industry itself, and its practices must be inferred from those of some other branches of the defence industry about which more is known. It is also the case that the actual relationships between institutions may not correspond to the formal, constitutional arrangements. To some extent this informal dimension can be grasped with the aid of Soviet historical works and the memoirs of participants, and also the evidence of émigrés, but there is an inevitable time lag rendering it difficult to appreciate the current situation. Moreover, it would be an error to believe that even the intelligence agencies of the West with their considerable human and technical resources have anything like a full appreciation of the realities of Soviet decision-making for national security.

> 'Some small chinks in the Soviet secrecy tradition have been opened in the course of SALT – such as agreement on exchange of a common data base, which may have appeared to the Russians a more momentous concession than to others from a different political culture. Anecdotally, for example, a Soviet official is said to have remarked: "There goes 500 years of Russian history" '.[3]

THE COMMUNIST PARTY OF THE SOVIET UNION

The Soviet Union is a single party state. According to the 1977 Constitution, the Communist Party is the 'leading and guiding force in Soviet society and the nucleus of its political system . . . armed with Marxism–Leninism [it] determines the general perspectives of the development of society and the course of the domestic and foreign policy of the USSR'.[4] Membership of the Communist Party is in excess of 18 million, drawn from all sections of society. Approximately one in nine of the adult population is a Party member. The organisational structure of the Party is determined by the principle of democratic centralism, according to which all leading bodies from the lowest to the highest are elected, strict discipline is required with the subordination of the minority to the majority, and the decisions of higher bodies are binding on those beneath. It is a hierarchical organisation, extending

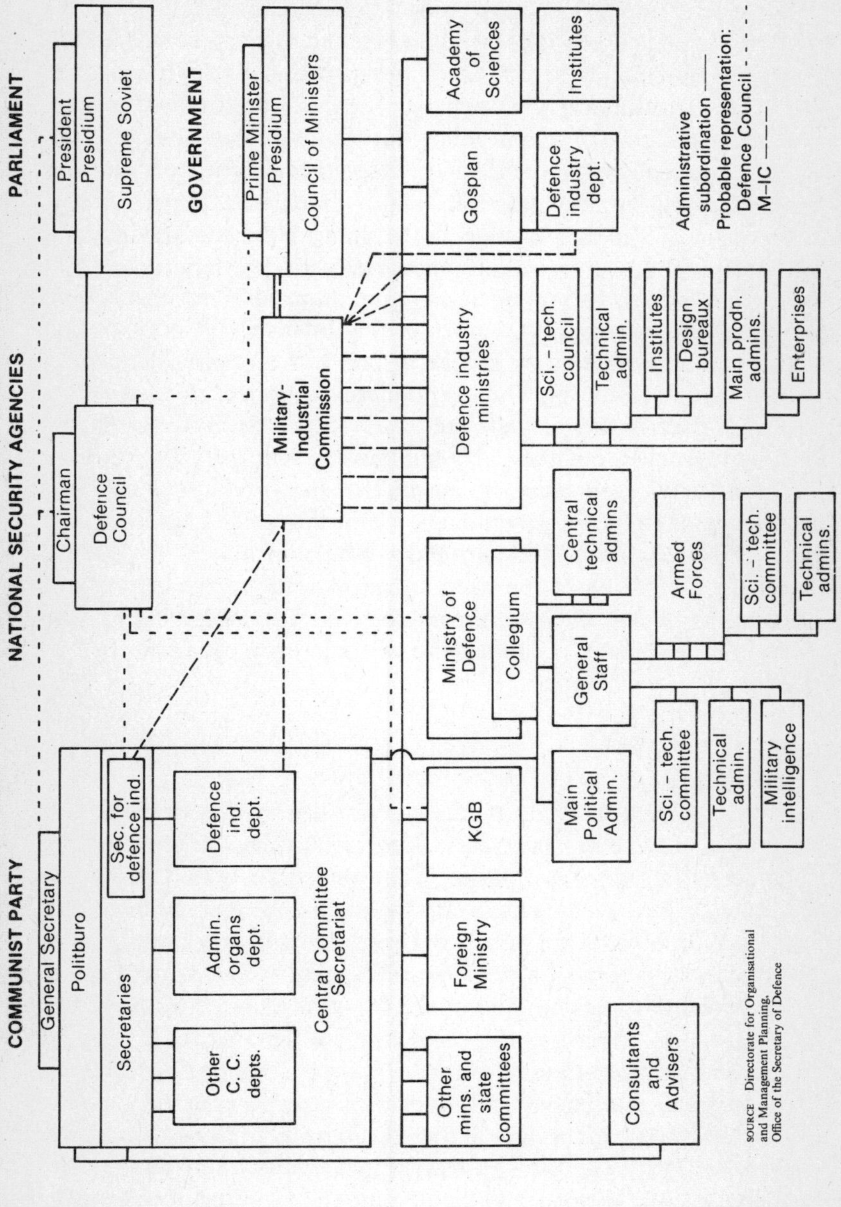

SOURCE Directorate for Organisational and Management Planning, Office of the Secretary of Defence

FIGURE 1.1 *The institutional framework of Soviet national security decision-making.*

from the lower, primary branches, some 450 000 in number, usually organised at places of work, to the higher, leading organs, the Politburo (political bureau), Secretariat and Central Committee.

The Communist Party is constitutionally and organisationally separate from the system of government administration, but in practice the lines of demarcation are often blurred. The Party determines the main lines of domestic and foreign policy, mobilises support for the fulfilment of these policies, and checks on the implementation of measures to put the policies into practice. It also exercises control over the appointment of leading personnel in the state administrative structure. The system of government is concerned with legislation, and the execution of policy. Its top officials are simultaneously members of the leading organs of the Party and are thus involved on both policy-making and implementation. Unlike Britain with its politically neutral civil service, the USSR has a highly politicised system of state administration: all government officials are expected to actively support and promote the policies of the Party and government departments have their own Party organisations to ensure that this occurs. This intimate relationship of Party and government means that some high-level state bodies effectively function as joint Party-government agencies, and the formal lines of demarcation are further blurred by the practice of adopting joint Party and government decrees on major issues of policy relating to the spheres of competence of state organisations.[5]

The Party Congress

According to its rules, the supreme body of the Party is the Congress which takes place every five years. Some 5000 delegates elected from constituent Party organisations meet for almost a week to review the experience of the preceding five years and discuss the main priorities for the period ahead. Foreign policy and security issues are considered in broad terms in the report of the General Secretary. Congress discusses and approves the draft outlines of the five-year plan for the economy and in so far as it incorporates projections of the growth of national income, industrial output, investment and consumption, it must be based on some assessment of the level of defence expenditure considered appropriate in the light of the prevailing international situation. At the 26th Party Congress in 1981 approximately 300 of the 5000 delegates had military rank and there is some evidence that the group of military delegates meets separately with the Party leadership at some point during the Congress week. Among the delegates are scientists, engineers, managers and shop-floor workers from organisations of the defence industry. The Congress selects the Party Central Committee, by secret ballot, but according to a single recommended list covering all the places on the Committee. This list is probably compiled in advance of the Congress by the Central Committee Secretariat.

The Central Committee

In considering the Party Central Committee a distinction has to be drawn between the Committee itself, which usually meets in full session twice a year, and the administration of the Central Committee with its substantial full-time Secretariat. The plenary sessions of the Central Committee discuss major issues of policy which arise between congresses, review the draft annual plans and elect the members of the Politburo and Secretariat, and the General Secretary. The Committee elected in 1981 consisted of 319 full members with voting rights and 151 candidate members with the right to speak but not vote. Of the total membership, thirty-six (8%) were uniformed members of the armed forces. The military members normally include the Minister of Defence and all his deputies, leading members of the General Staff and some commanders of military districts. However, the total defence representation is somewhat larger as all the ministers and leading Party and government officials responsible for weapons production are usually members, and also some scientists, designers and managers associated with the armaments industry. It can be estimated that in 1981 the total defence representation among full members was some 12%; for comparison, the foreign affairs group accounted for approximately 5% of the membership. In terms of policy-making these proportions may not have much significance, but there is no doubt that membership of the Central Committee is a matter of some prestige and must facilitate access to the real centres of power. For those in the industrial and research system, there may also be practical benefit in terms of an enhanced ability to command resources and to resolve problems by direct resort to the highest political authorities.

The Politburo

The real centre of political decision-making in the USSR is the

Politburo, usually consisting of twenty to twenty-four members under the chairmanship of the Party General Secretary. Following the pattern of the Central Committee that elects it, the Politburo has full and candidate members. Like the British Cabinet, it meets every Thursday; unlike the Cabinet, an official report of its deliberations is published in the national press on the following day, a practice introduced under Andropov. In recent years its members have included leading Party officials from the Secretariat and First Secretaries of a number of Republican and city Party Committees, and also the heads of a number of the most powerful agencies of the state, including government departments, the armed forces, the foreign ministry and the state security organisation. Other ministers and officials also attend Politburo meetings when particular issues relating to their competence are under discussion.

The agenda for the Politburo is prepared by the Party Secretariat and given the overlapping membership of the two bodies it is likely that many issues have been effectively resolved and decisions drafted in advance of the actual Politburo meeting. Some evidence suggests that this applies more to domestic matters than to foreign policy questions, on which the Politburo may well have a more substantial independent decision-making role. From the published accounts of its deliberations it is clear that the Politburo establishes priorities, takes decisions and exercises oversight over the full range of domestic and international affairs. One authoritative Soviet work states that 'Military development is at the centre of attention of the Politburo of the Communist Party Soviet Union (CPSU), which adopts resolutions concerning both the defence of the country as a whole, and also concrete measures directed towards the improvement and development of the Armed Forces'.[6]

Within the collective leadership of the Politburo the Party General Secretary is the dominant figure. Whereas other Party Secretaries have certain broadly defined spheres of competence, the General Secretary's brief extends to the entire range of issues facing the Soviet state and in his work he is aided by a number of personal assistants whose role appears to have increased in recent years.

The Central Committee Secretariat

So far we have considered leading Party bodies having general responsibility for political leadership of the country, including matters of national security. The Secretariat is the principal administrative organ of the party and has within it departments and officials with specific responsibility for questions relating to the armed forces and the development and production of weapons. According to the Party rules, the Secretariat is responsible to the Central Committee for directing 'current work, chiefly the selection of cadres and the organisation of the verification of the fulfilment of decisions'.[7] The Secretaries, usually about ten in number, are elected by the Central Committee and oversee particular areas of work and the departments associated with them. The responsibilities of one of the Secretaries, currently Romanov, include oversight of the defence industry and, possibly, some aspects of the work of the armed forces and the security services.

> During the late 1950s when the Soviet ICBM force was under development Brezhnev was the Party Central Committee Secretary for the defence industry: 'During those years the office of the Secretary of the CC was a kind of staff headquarters where the most important problems of missile technology were resolved, and meetings held with the participation of the most eminent scientists, designers and specialists in various fields of science, technology and production. L. I. Brezhnev was often seen in the factories where missile technology was being created.'[8]

Ustinov, who occupied this position between 1965 and 1976, before becoming Minister of Defence, is said to have 'co-ordinated and directed the activity of scientific establishments, design bureaus and industrial enterprises for fulfilling the assignments of the Communist Party for the further consolidation of the economic and military might of the motherland'.[9] There is a special Defence Industry Department of the Central Committee, which monitors all weapons development and production (and the space programme), and has sub-departments concerned with particular branches of the defence industry, including, presumably, nuclear weapons. There is also a Department of Administrative Organs, about which little is known, but it appears to exercise control over senior appointments in the armed forces and the KGB.

However, the Secretariat does not appear to have a department which monitors the routine work of the Ministry of Defence in the same way that other activities are controlled. The Main Political

Administration of the Armed Forces is not part of the Central Committee Secretariat, but has the rights of a Central Committee department. This administration leads all Party and political work in the forces with concern for such questions as political education, ideological and patriotic commitment, and the strengthening of military discipline and preparedness.

The Secretaries meet on a regular basis together with the heads of departments to review matters of current concern and prepare materials for the Politburo. These meetings are often attended by the editors of the principal Party and government newspapers, the head of the radio and television service, academic foreign policy specialists, the President of the Academy of Sciences and other leading figures who participate in the discussion. Their lack of voting rights is apparently of little importance as it appears that votes are rarely taken.[10] Thus, despite the existence of a formal organisational structure and rules, the actual functioning of the highest level Party bodies is less rigid than might appear at first sight.

An important aspect of the Central Committee's work is its responsibility for the selection and placement of leading personnel. Many responsible posts within the institutions of the state are covered by the *nomenklatura* system, according to which those occupying them must be on a list approved by the central Party authorities. This is one of the crucial mechanisms by which the Party exercises its leading and guiding role in Soviet society. Another important mechanism is the adoption by the Central Committee of resolutions outlining measures to be implemented by lower Party organisations to tackle problems identified by the centre. These resolutions are drafted by the Central Committee departments and the Secretariat and passed to the Politburo for approval. On significant questions concerning the work of state institutions the Party will often secure the adoption of decrees in the name of both the Central Committee and the Council of Ministers – the highest executive government body, described below. Whereas the Party resolutions are binding only for other Party organisations, the joint decrees have the force of law.

The structure of Party organs outlined above applies to the USSR as a whole. Each of the Republics, except the largest, the Russian Socialist Federation of Soviet Republics (RSFSR), has its own Party Central Committee, Politburo and Secretariat, and there are leading committees in the regions and cities. The Republics, and some regions and cities have their own Party defence industry departments and it is likely that they maintain close links with the central department in Moscow, keeping it informed of local problems.

THE STATE

Questions of national security and nuclear weapons in particular are handled by a number of institutions of the Soviet state. These include the Supreme Soviet (the Soviet Parliament), the system of government (the Council of Ministers), the state security organisation (the KGB), and various ministries and government departments, important among which are those involved in the actual development and production of weapons systems.

The Supreme Soviet

According to the Constitution, the highest body of state authority competent to take decisions on all issues of national importance is the USSR Supreme Soviet. It is the Soviet Parliament with its elected deputies that enacts laws, approves state plans for economic and social development, and forms the Council of Ministers. The Supreme Soviet has two chambers, having equal rights, and a number of Standing Commissions, which can examine the work of government bodies and initiate and prepare new legislation. One of the Standing Commissions is concerned with international affairs, but there is no commission for defence matters. The Supreme Soviet normally meets only twice a year; continuity is provided by the Standing Commissions and, above all, the Presidium of the Supreme Soviet.

The Presidium of the Supreme Soviet

The Presidium, which has approximately forty members, is elected from the deputies of the Supreme Soviet; its Chairman is the Soviet President and in recent years the position of president has been occupied by the Party General Secretary. The formal powers granted to the Presidium include the ratification of international treaties and responsibility for proclaiming a state of war. The Constitution also states that the Presidium of the Supreme Soviet shall 'form the Council of Defence of the USSR and confirm its composition; appoint and dismiss the high command of the Armed Forces of the USSR'.[11]

The Defence Council

The existence of the Defence Council was not publicly revealed until 1976, when it was disclosed that Brezhnev was its chairman. Later it became known that he had occupied this post since his election as General Secretary in 1964. In the West there has been speculation as to the status of the Council, its membership and functions, with claims that it is a Party body, in effect a sub-group of the Politburo. Recent research casts doubt on this interpretation: at least from a formal constitutional point of view the Defence Council is an organ of state administration, the supreme state body with competence in military affairs.[12] Its status is probably that of a committee of the Presidium of the USSR Supreme Soviet and, as such, it will have the authority to co-ordinate the affairs of all state agencies concerned with national security, including the Council of Ministers, and to take legally binding decisions. The latter would not be true if the Council were purely a Party body. As a state organ of the very highest level one would expect its membership to consist predominantly of members of the Politburo in so far as they head the principal security-related state organs. In practice it may well be the case that the Defence Council is effectively a joint Party and government body. According to a Soviet work on administrative law, the Defence Council is headed by the Chairman of the Presidium of the Supreme Soviet: a fact which may help to explain why this post has been taken by the Party General Secretary since the new Constitution was adopted in 1977.[13] In this capacity the Party leader and President can be considered the Commander-in-Chief of the Armed Forces. Other likely members are the Minister of Defence, the chairman of the Council of Ministers, and the head of the KGB. Possible candidates also include the Foreign Minister and the Party Secretaries concerned with the defence industry and foreign affairs. Following the practice of the Politburo and Secretariat it is likely that other officials attend its meetings when required.

The Defence Council has wide powers with respect to all aspects of national security. From fragmentary evidence in Soviet sources it can be deduced that it co-ordinates the defence-related activities of state and government agencies, reviews and decides all basic questions relating to the country's 'military development' and has an important role in arms control decision-making. It may also oversee and co-ordinate the work of the KGB and the intelligence services. The Soviet term 'military development' refers to the entire system of economic, political and strictly military measures adopted by the state to enhance its military potential. It therefore concerns such questions as the structure and organisation of the armed forces, military personnel and technical policy, and the procurement of weapons. It is likely that the Defence Council oversees the weapons acquisition process, takes decisions on specific systems, and provides overall direction and co-ordination of the work of the various agencies concerned, including the Military–Industrial Commission and the armaments administrations of the Ministry of Defence and the individual services. In the event of war, the Defence Council would probably be converted rapidly into the supreme organ of state administration, similar to the State Committee for Defence which led the USSR during the last war.

The Council of Ministers

The Council of Ministers is the Government of the USSR, in the words of the Constitution, 'the highest executive and administrative body of state authority'. It is formed by the Supreme Soviet and consists of a chairman (the Prime Minister), first deputy chairmen (currently three), and deputy chairmen (eleven), who together constitute its Presidium, and the ministers and chairmen of state committees, plus the chairmen of the Republican Councils of Ministers, making a total of some 130 members. It has wide powers to deal with matters of state administration apart from those strictly within the competence of the Supreme Soviet and its Presidium. All the national industrial ministries, the state committees concerned with planning, science and technology, and other economic matters, the Academy of Sciences, the Ministry of Defence and the KGB are subordinate to the Council of Ministers, the decrees and orders of which are legally binding. The Council is charged with ensuring the direction of economic, social and cultural development, and with exercising general direction of the development of the armed forces. The full Council meets occasionally, but its effective standing body is its Presidium, which, like the Politburo, meets on a weekly basis. The membership of the Presidium includes some of the most powerful (state) officials, the heads of the Foreign Ministry, the State Committees for Planning, Science and Technology and Supply, and of the Military–Industrial and Agricultural–Industrial Commissions. It does not include representatives of the military or KGB, presumably because they have their own higher-level state body, the Defence Council.

The Military–Industrial Commission

The Military–Industrial Commission (VPK – *voenno-promyshlennaya komissiya*) of the Presidium of the USSR Council of Ministers is a body whose existence is very rarely acknowledged in Soviet publications. A Commission with the same title was first established in January 1938, charged with the task of mobilising and preparing industry, both military and civilian, for the fulfilment of plans for the production and delivery of weapons to the amed forces.[14] Little is known with certainty about the present-day VPK. The ministers of the nine defence industry ministries, described later, will certainly be members of the Commission, which is presided over by a Deputy Chairman of the Council of 12 Ministers (L. V. Smirnov). Other members probably include the Party Central Committee Secretary responsible for the defence industry and the head of the Defence Industry Department of the Central Committee; indeed, there is evidence that the former chairs some of its meetings.[15] Thus, in practice, like the Defence Council, the VPK should probably be considered a joint Party and government body. Other ministers are likely to attend when matters concerning their competence are under discussion. Following the example of another body of similar status, the Agricultural–Industrial Commission established in 1982, the VPK can be expected to have the following functions:

(1) co-ordination of the activities of the ministries concerned with military production, including the defence-related work of civilian ministries;
(2) general oversight and checking of the fulfilment of Party and government decisions and the assignments of state plans;
(3) preliminary consideration of draft plans for the development of the defence industry prepared by the State Planning Committee (Gosplan); and
(4) preparation of proposals for action on specific issues for submission to the Defence Council.

In addition, the VPK appears to have major responsibilities in relation to weapons research and development, including project selection and approval, and the co-ordination of programmes involving organisations of different ministries. There is evidence that it also has a central role in the system of scientific and technical information serving the defence sector, acting as a clearing house for material gathered by the intelligence services and other agencies. For the civilian economy these latter two functions are fulfilled by the State Committee for Science and Technology; in this respect, the VPK can be considered a defence-related equivalent of the State Committee, and, as such, can be expected to have a sizeable permanent apparatus and staff.

As a Commission of the Presidium of the Council of Ministers, the VPK has a higher status than the ministries subordinate to the Council, and this must endow it with considerable authority and provide an effective mechanism for resolving inter-departmental disputes. Until the formation of the Agricultural–Industrial Commission, there was no equivalent organisation for any sector of the civilian economy, which suffers from the absence of agencies for effective high-level co-ordination.

The State Planning Committee (Gosplan) and other State Economic Agencies

As the body responsible for economic planning in the Soviet Union, Gosplan must inevitably have a role in the weapons procurement process and national security decision-making. The development of the defence sector is planned like the rest of the economy and it is believed to be Gosplan's task to ensure that the necessary resources are available to fulfil the programmes undertaken by the defence industry ministries. It has a special defence industry department and one of its first deputy chairmen leads the military–economic side of Gosplan's work. This official probably participates in the deliberations of the Military–Industrial Commission, and can be assumed to have close relations with the Defence Industry Department of the Central Committee. Testimony to the closeness of these three bodies is the fact that over the years a number of senior officials have transferred from one to another. Details are not available, but it is likely that other state economic bodies also have special military departments, for example, the Ministry of Finance, and the State Committees for Supply, Labour, Prices and Standards.

The Committee for State Security (KGB)

The KGB has wide involvement in issues of national security. It is a secret police and intelligence service and has responsibility for a section of the armed forces, the Border Guards. Its special troops control the movement of and stockpiles of nuclear weapons, guard strategic installations and control communications among the high-level Party officials and the upper ranks of the military. The KGB also has responsibility for maintaining secrecy and security in organisations connected with weapons development and production. Research institutes and enterprises engaged in military work have so-called First Departments, overseen by the KGB, which control access to classified documents and ensure the security of contacts with other organisations. This activity is supplemented by a system of security vetting for personnel. Another role of the KGB is the gathering of foreign, scientific and technical intelligence, used by the military and industry to monitor Western developments and as a source of ideas applicable to the domestic weapons development effort. Technical intelligence is also undertaken by other state agencies, including the military intelligence service, the GRU.

The Ministries of the Defence Industry

The Soviet defence industry represents an important sector of the socialist planned economy.[16] According to the US Department of Defence, more than 150 major final-assembly plants build weapons as end products, supported by thousands of factories supplying components and other inputs.[17] The size of the defence industry labour force is not known, but must run into several million. It can be tentatively estimated that some 50–60% of qualified scientists and engineers serving industry work for the defence sector. In many respects the Soviet defence industry is similar to its civilian counterpart. All enterprises and facilities of Soviet industry are State-owned and administered by ministries subordinated to the USSR Council of Ministers. These ministries are not unlike the State corporations that run nationalised industries in Britain: most of Soviet industry is in the hands of some three dozen equivalents of the National Coal Board. The following nine ministries are generally considered to form the core of the defence industry:

Defence Industry:	ground forces equipment – tanks, other armoured fighting vehicles, artillery, rocket launchers, small arms, and optical equipment;
Machinebuilding:	conventional ammunition and explosives;
Aviation Industry:	military and civilian fixed-wing aircraft and helicopters;

Shipbuilding Industry:	military and civilian ships and boats;
Electronics Industry:	electronic components, micro-computers and some other electronic end-products;
Radio Industry:	radio and radar equipment;
Communications Equipment Industry:	radio, telephone and telegraph communications equipment, including satellite communication systems;
General Machinebuilding:	strategic missiles, rockets and space vehicles;
Medium Machinebuilding:	nuclear warheads and devices, uranium mining and processing.

Some civilian ministries produce weapons and other military-related products; many defence industry enterprises produce civilian goods, for example, the Ministry of General Machinebuilding manufactures tractors, machine tools, trams, refrigerators, televisions and medical equipment. All the ministers of the defence industry can be expected to be members of the Central Committee.

The industrial ministries have a common internal structure. At the level of the minister there is a Collegium, a consultative body of the senior officials which reviews basic issues of policy and makes recommendations implemented through ministerial orders. At this level there is also a Scientific and Technical Council, consisting of the leading scientists, designers and engineers of the industry. The Council examines research proposals and new technical projects, and attempts to secure a coherent technical policy for the ministry. Under the minister are a number of functional administrations – for planning, financial matters, labour, supply and other areas, including a main inspectorate for quality. For weapons development, the principal body is the Technical Administration, responsible for the research and development effort of the ministry, and to which are subordinated the most important, central research institutes and design bureaus. The ministry also has a number of main Production Administrations, specialised by type of product, which control the manufacturing enterprises. In recent years many enterprises have been amalgmated to form production associations, or science-production associations, the latter uniting research, design and manufacturing facilities. The activities of the functional and production administrations are overseen by deputy ministers, usually six to eight per ministry.

Like its civilian counterparts, a defence industry enterprise operates according to a plan setting out certain basic targets and is expected to economise on resources and achieve a profit, part of which is used to pay bonuses to its workforce. There are special features of the defence sector, however, which mark it off from the rest of the economy and create conditions for its successful functioning. It enjoys first priority in the supply of materials in terms of both quantity and quality. There are also special privileges for workers, and managerial and technical personnel, in the form of higher pay scales and bonuses, and better-than-average housing and welfare services. Against these benefits must be set the very high degree of secrecy and some limitations on career advancement and public recognition. Designers, engineers and managers in the weapons industries are severely restricted in their ability to travel abroad and openly publish, and may not receive public acknowledgement of their contribution until after their retirement or death.

The Ministries for Nuclear Weapons and Missiles

Of all the Soviet industrial ministries the one about which least is known is the Ministry of Medium Machinebuilding (Minsredmash), responsible for all nuclear weapons development and production.[18] The ministry as such was established in 1953. Prior to this, the nuclear weapons programme had been managed by a body called the First Main Administration of the USSR Council of Ministers, created in 1946, with overall control in the hands of Beria, the head of state security. This ministry has had remarkable continuity of leadership: since 1957 its minister has been E. P. Slavskii, born in 1898 and associated with the nuclear programme almost from its inception. The ministry has responsibility for all stages of the production of nuclear warheads from the extraction and processing of uranium ores to the manufacture of nuclear devices. Uranium mining is quite widely dispersed in the USSR, with major centres in the Ukraine, and the Central Asian republics. The Chelyaninsk region in the Urals appears to be the principal centre for the fabrication of nuclear warheads and associated research and design. But Minsredmash also plays a major role in the civilian nuclear power programme: it has responsibility for the entire fuel cycle, with the exception of the commercial power stations themselves, which now come under the civilian Ministry of Power and Electrification.

In 1956, with the development of civilian applications of atomic energy, part of the nuclear industry was transferred to a newly created State Committee for the Utilisation of Atomic Energy. This move permitted a lowering of the barriers of secrecy and facilitated the extension of international contacts in the nuclear field. Today, many of the leading nuclear research institutes are subordinated to this State Committee, including the Kurchatov Institute of Atomic Energy in Moscow, the principal Soviet nuclear research centre, under the directorship of Academician A. P. Aleksandrov, President of the USSR Academy of Sciences. Given the scope of the work of some of the institutes, for example, research into plutonium and other nuclear materials undertaken by the Moscow All-Union Research Institute for Inorganic Materials, it is likely that they contribute to the activity of Minsredmash, and it is possible that some of them are under dual control of both the State Committee and the ministry.

There are two major centres for underground testing of nuclear devices. The oldest is located in the Semipalatinsk region in the deserts of eastern Kazakhstan. The second centre is on the island of Novaya Zemlya. However, the Soviet Union also uses nuclear explosions for civil engineering projects, including the re-creation of underground natural gas storage chambers.

Created in 1965, the Ministry of General Machinebuilding (Minobshchemash) is responsible for the development and production of ballistic missiles and space vehicles. The ministry has four Main Production Administrations – for missiles and space rockets (and, probably, other space vehicles), missile engines and motors, guidance systems and ground equipment. According to the US Department of Defence, there are more than twenty plants building missile systems,[19] supported by a network of enterprises supplying engines and other components. During the early 1950s there were two design bureaus for ballistic missiles, named after their prominent designers – the Korolev bureau at Kalingrad near Moscow and the Yangel bureau in or near Dnepropetrovsk in the Ukraine. Later in the 1960s the former appears to have transferred exclusively to space rockets and vehicles, but by then two new *opytno-konstruktorskoe byuro* (OKB) had been created: the Chelomei bureau for ICBMs, submarine-launched missiles, armed cruise missiles and the Nadiradze bureau for solid-fuel missiles, both apparently located in or near Moscow.[20] Some sources suggest that development work on sea-based missiles may have been transferred from the Chelomei bureau to a new organisation so that there are currently four strategic missile design centres. There are also leading research and design organisations for engines and guidance systems.

Minobshchemash has a system of central research institutes, although few have been identified. The largest appears to be NII-88 at Kalingrad, which has its origins in the Jet-Propulsion Research Institute (RNII) created in 1933. Following the pattern of the aircraft industry, of which the missile industry was formerly part, the principal OKB appear to have pilot production facilities for building prototypes of new missiles and engines, which are then transferred to series production plants.

The Soviet Union has three large rocket launch and test centres, or 'cosmodromes' as they are termed, at Kapustin Yar near Volgograd, Plesetsk in the northern Archangel region, and Baikonur (Tyuratam) in Kazakhstan, the latter being the launch centre for manned space flights. Overall control of these centres appears to be in the hands of the military. It is from these sites that new missiles are test-launched.

Weapons Research and Design Organisations

Most applied research and development work on weapons systems is undertaken in the institutes and design bureaus of the defence industry ministries. There is no standard organisational form for these establishments. Some institutes have attached design and experimental production facilities; others have been incorporated in science-production associations. One important form is the experimental-design bureau, (OKB) especially well-developed in the aviation industry, but probably also in the missile industry and other branches. Here the design office is linked with an experimental production factory and test facilities, enabling it to transfer fully-proven new designs for manufacture at the series production enterprises of the ministry. A large OKB undertaking work on a number of different types of aircraft or missiles and is headed by a General Designer, who leads the work of Chief Designers responsible for particular new systems. Within the OKB the General Designer chairs sessions of its Scientific and Technical Council, a consultative organ involving the leading designers and specialists of the organisation. Here new projects have to be reviewed to establish their viability and scientific value, and reports are heard on the fulfilment of research and design programmes. The General Designers have substantial authority and must be regarded as major figures in the weapons development process. Some of the leading missile and aircraft General Designers are members of the USSR Academy of Sciences, the Party Central Committee and the Supreme Soviet. From Soviet memoirs and other historical evidence it is clear that leading designers may have direct access to the top political leadership and it is likely that they participate from time to time in the Military–Industrial Commission and, possibly, the Defence Council.

The Academy of Sciences and the Higher Education System

Most of the country's basic research and some applied research in high-technology fields is undertaken by institutes of the USSR Academy of Sciences and the Republican Academies. Although Academy institutes do not appear to work on weapons as such, some undoubtedly do work of actual or potential military relevance. Indeed, there is evidence that this involvement in military-related R & D has grown in recent years. Such research includes work in the fields of nuclear physics, lasers, electronics, new materials, computers, control systems and advanced manufacturing technologies. Election to the Academy is considered a high honour and, as noted above, members include leading designers, scientists and engineers directly associated with weapons development. The Academy president, Aleksandrov, and some Vice-Presidents and other Academicians are members of the Central Committee.

Research for the defence industry is also undertaken, often on a contract basis, by technical institutes of the higher education system. Some of these have long-standing, close relations with the defence sector and supply it with many qualified graduate engineers and scientists, for example, the Moscow Bauman Higher Technical Institute and the Moscow Aviation Institute with the aerospace industry, and the Moscow Physical–Technical Institute with the nuclear industry.

Consultants, Advisers and Citizens

A feature of the policy process in the USSR in recent years has been the increasing role of specialist advisers. This has been particularly evident in relation to foreign policy, but also applies in other fields. Leading staff of such Academy of Sciences establishments as the Institute for the World Economy and International Relations, and the Institute for the Study of the USA and Canada, prepare reports at the request of the Central Committee and may attend sessions of the Secretariat and Politburo. There may be similar, less publicised, inputs from specialists working in the leading military academies, including the Academy of the General Staff, the Frunze Academy and the Lenin Military Political Academy. It is doubtful whether such advice is of decisive importance, but it probably ensures that the real decision-makers are better informed of the policy options and their implications.

Issues of foreign policy and national security are discussed in the Soviet press, including the military newspaper, *Krasnaya Zvezda* (Red Star), and in more specialised journals available to the interested public, although the principal military, theoretical journal, *Voennaya Mysl'* (Military Thought) is not openly distributed. There are also various fora involving natural and social scientists, including the Scientific Council for the Study of Problems of Peace and Disarmament sponsored by the Academy of Sciences, the Committee of Soviet Scientists in Defence of Peace, Against Nuclear War and the Soviet Pugwash Committee. Interested citizens can participate in the activities of the Soviet Peace Committee and express their views through their Supreme Soviet delegates and letters to the Central Committee. The latter has a special Department of Letters, one of a range of means by which the central leadership monitors public opinion. However, while the concerned Soviet citizen can now be reasonably well-informed of the broad issues of national security, the crucial institutions and mechanisms for decision-making in relation to nuclear weapons remain veiled in almost total secrecy.

THE MILITARY

The armed forces of the Soviet Union have a highly centralised command structure. The primary administrative organ is the Ministry of Defence, subordinate to the Council of Ministers, and headed by a minister, three first deputies and eleven deputies. One of the first deputy ministers is Chief of the General Staff, another the Commander-in-Chief of the Warsaw Pact Forces. Although not formally of this status, the head of the Main Political Administration of the Armed Forces mentioned earlier would seem to be the equivalent of a first deputy minister; the Administration probably operates under dual subordination to both the Ministry of Defence and the Party Central Committee. All these top officers are normally members of the Central Committee. The ministry has a high-level consultative body, the Collegium (or Main Military Councils, as it used to be called), chaired by the Minister, which considers questions relating to the development of the forces, their military and political training and personnel policy. Its recommendations are put into force through orders of the ministry. A significant feature of the Soviet Ministry of Defence is that it does not have any high-level civilian participation. As an entirely uniformed, professional military body it differs from both the British Ministry of Defence and the US Department of Defence. The principal organ of management of the forces is the General Staff.

The General Staff

The General Staff is subordinated directly to the Minister of Defence. It co-ordinates the activities of the various service branches, the Rear Services (food and clothing supply, medical services, accommodation, transport, etc.), Civil Defence and the main administrations of the ministry. According to the Soviet Military Encylopaedia, the General Staff:

> thoroughly analyses and evaluates the evolving military–political situation, determines the trends of development of the means of waging war and the methods of their use, organises the training of the armed forces and carries out necessary measures for securing their high combat readiness to repulse any possible aggression. The further development of military theory occupies an important place in the activity of the General Staff. It directs military scientific work, elaborates the most important regulations and topical problems of Soviet military science, and introduces the achievements of the latter into the practice of operational and combat training of troops and staff.[21]

To undertake this work it has a Military Science Administration and also a Scientific and Technical Committee, which probably analyses the development of weaponry, clarifies issues of technical and procurement policy, and helps to identify new weapons requirements. Also located within the General Staff is the Military Intelligence Administration (the GRU) which gathers information on potential adversaries and assists the General Staff in preparing threat assessments which help to determine force requirements and procurement needs. In addition the General Staff has administrations for operations, foreign military assistance, communications and external relations.

All the evidence indicates that the General Staff is an extremely important body in national security decision-making. It is probably the major source of information for the Defence Council, and represents a mediating body between the top political leadership and the forces. Not only does it elaborate fundamental questions of military strategy and doctrine, but it also acts as a unified operational command centre for the five services. Of all the agencies of the Soviet state, the General Staff must be the best informed on foreign military–technical developments and this information probably serves as an important input into the weapons acquisition process. Given the scope of its activities it is not surprising that the General Staff has played an active role in the arms control process, providing military representation on the SALT delegations. Negotiating positions are probably clarified within the General Staff before being presented to the Defence Council and Politburo for approval.

At the Brezhnev–Nixon summit meeting of July 1974 Soviet General Staff officers gave a briefing to the American side: 'The Americans were disheartened. One of those present said that he thought that Brezhnev had been "sold a bill of goods" by the Soviet military. Subsequently, the Americans checked the Soviet assessment and discovered that it was technically justified, provided all the optimistic statements made by American admirals and generals about the new weapons were taken at face value and interpreted at what would be from the Russian point of view the worst possible fashion'.[22]

Following the precedent of 1941–5, in the event of war the General Staff together with the Party General Secretary, as Commander-in-Chief of the Armed Forces, would constitute the headquarters of the Supreme High Command for military leadership of the war effort.

Central Technical Administrations of the Ministry of Defence

One of the Deputy Ministers of Defence has responsibility for armaments and as such leads the weapons acquisition process from the military side. During some periods, but not at present, there has been an additional Deputy Minister for Electronics. The Deputy Minister for Armaments leads the work of a number of administrations concerned with the procurement of specific types of weapons. These include the Main Rocket and Artillery Administration (probably limited-range battlefield rockets only), and the Main Armour Administration (tanks and armoured vehicles). Details are not available, but there are probably similar administrations for nuclear weapons and longer-range ballistic missiles. These bodies can be expected to play an active role in the procurement process, with duties including the identification of new weapons requirements, project selection and approval, and the acceptance of newly developed systems. Another central administration of the ministry leads the work of the military educational establishments, some of which undertake weapons-related research and analysis.

The Armed Forces

The five principal service branches are subordinate to the General Staff; their Commanders-in-Chief are Deputy Ministers of Defence. The services are the Ground Forces, the Air Defence Forces, the Air Force, the Navy and the Strategic Rocket Forces, the latter created in 1959.

The Soviet nuclear capability is shared between the different services. The principal nuclear strike force of intercontinental and intermediate range missiles is under the Strategic Rocket Forces; the Navy controls the nuclear-armed submarines and tactical nuclear weapons carried by ships and naval aviation; the Air Force controls the long-range nuclear bomber force and nuclear-capable tactical air forces; and the Ground Forces possess nuclear missiles of shorter range and nuclear artillery. The Air Defence Forces control the early-warning system and the anti-ballistic missile system around Moscow; the interceptor missiles of the latter carry nuclear warheads.

In each of the services there appears to be a Deputy Commander-in-Chief responsible for armaments and questions of procurement. They probably head special administrations which, together with the central technical administrations of the Ministry of Defence, constitute the 'customer' for the weapons produced by the defence industry. Their responsibilities include the issuing of requirements for new weapons, general oversight of the R & D process, the acceptance testing of prototypes of new weapons, and probably leadership of some weapons-related investigations undertaken by their own research institutes. The services also appear to have their own Scientific and Technical Committees, which probably elaborate issues of technical policy and review proposals for new weapons development. It is also the various technical administrations of the ministry and the services that provide the military representatives (*voenpredy*) who monitor the work of research organisations and enterprises to see that specifications and quality standards are observed. Finally, the services have their own military academies and schools, some of which appear to be important not only for their officer training, but also as centres of research, for example, the two military academies of the Strategic Rocket Forces in Leningrad and Moscow and the Moscow Zhukovskii Military Air Engineering Academy of the Air Force.

THE WEAPONS DEVELOPMENT PROCESS

The procedure for developing new weapons has clearly defined stages and decision points and in general terms is the same for all types of technology.[23] It is illustrated in Figure 1.2. The basic forms of technical documentation and the content of the various stages are regulated by obligatory state standards. What does seem to vary is the level within the industrial, military and political hierarchies at which decisions are made and this clearly depends to a great extent on the character of the system, its military significance and political sensitivity.

The initiative for a new project can arise either from a design bureau or research institute of industry, or from the military customer who identifies a requirement for a new system. The design bureaus of the aviation industry, and presumably other branches, have departments concerned with the investigation of future design possibilities and they may make proposals which are first reviewed by the design bureaus' own Scientific and Technical Council and then the equivalent Council of the Industrial Ministry. On the military side the initiative may come from a Technical Administration or Scientific and Technical Committee of one of the services, or possibly the General Staff, which identifies a need to match the technical innovations of a potential adversary or to meet new demands posed by changes in military doctrine. For new aircraft and other 'large technical systems' (as major weapons systems are termed in Soviet open literature) it appears to be normal practice for the design bureau, in consultation with the customer, to draw up a Preliminary Advance Project (*avanproekt*), setting out a basic design concept and certain desired performance and combat characteristics. This Advance Project is examined by the Scientific and Technical Council of the industrial ministry concerned,

STAGE OF DEVELOPMENT PROCESS	ORGANISATIONS INVOLVED
Project identification	Technical Administrations, Scientific–Technical Committees and Institutes of MoD, General Staff and Armed Forces; Industrial Design Bureaus and Institutes
Preparation of advance project	Industrial Design Bureau/Military Customer (Technical Administration of MoD or branch of Armed Forces)
Evaluation of advance project	Scientific–Technical Council of Industrial Ministry; Military Customer
Decision to proceed with project	Military–Industrial Commission; Defence Council; Politburo
Preparation of tactical–technical assignment	Military Customer/Design Bureau
Preparation of preliminary draft design	Design Bureau
Evaluation of preliminary draft design	Scientific–Technical Council of Design Bureau; Research Institutes of Industrial Ministry; Military Customer
Approval of preliminary draft design	Military Customer; Industrial Ministry
Preparation of technical project and working drawings	Design Bureau
Building of prototypes	Experimental Factory of Design Bureau
Testing of prototypes	State Commission, headed by Military Customer
Decision to transfer to series production	Military–Industrial Commission; Defence Council; Politburo
Series production	Manufacturing Enterprise of Industrial Ministry
Acceptance testing of first serially produced systems	Military Customer
Deployment of new system	Armed Forces

FIGURE 1.2 *The development process for a major new weapons system*

with the participation of representatives of the military customer. On the basis of the Council's deliberations a draft decree is prepared, outlining the new project, and identifying its General Designer and other executors, and also the sources of finance for the work. This is submitted for government approval, presumably to the Military–Industrial Commission or, possibly, the Defence Council (and Politburo) in the case of major systems. This results in a Council of Ministers decree initiating design work on the new weapons.

The next stage is the preparation by the Technical Administration of the service customer of an important document, the Tactical–Technical Assignment (TTZ–*taktiko-tekhnicheskoe zadanie*). This sets out in greater detail the technical, combat, operational and economic requirements for the new system, including an assessment of expected volume of production. This provides the basis for monitoring the work of the design bureau and assessing the final result. When the customer lacks the necessary knowledge and expertise the design organisation itself draws up the TTZ, which is then reviewed and agreed by the customer. For simpler weapons the Advance Project stage appears to be dispensed with and work begins immediately with the TTZ; in such cases it is likely that the project is approved at a lower level of the hierarchy.

On the basis of the TTZ the design bureau elaborates a Preliminary Draft Design (EP–*eskiznyi proekt*), a detailed exposition of how the assignment is to be fulfilled. This is reviewed by the Scientific and Technical Council and also submitted to the central research institutes of the ministry which have to certify that it meets required standards in terms of materials, aerodynamic qualities, serviceability and technological suitability for series production. The EP is submitted to the military customer for review and approval, and also the Industrial Ministry, which creates a special commission to examine and accept it. The approved EP provides the basis for drawing up a comprehensive Technical Project and this in turn is used for preparing detailed working drawings for building prototypes at the OKB's experimental production factory. These prototypes first undergo factory testing to see that they conform to the original TTZ and are then submitted for state trials at a military proving ground or test centre to assess their combat worthiness and confirm that the design specification has been fully met.

If the design is approved as suitable for series production, a further document, the Technical Conditions (TU–*tekhnicheskie usloviya*) is drawn up, setting out the specification and requirements for quality and delivery. This effectively constitutes a contract between the military technical administration and the ministry responsible for production. For major systems the initiation of series production appears to be sanctioned by a Council of Ministers decree and as such presumably has the approval of the Military Industrial Commission and, possibly, the Defence Council (and Politburo). Personnel from the OKB will assist in the assimilation of the series production of the new system, the first examples of which will have to undergo acceptance trials organised by the military before they are deployed by the forces. The entire development process is monitored by the military representatives permanently based at the design bureaus and enterprises, but subordinate to the military–technical administration. These arrangements must lead to the development of fairly close relations between the industrial personnel on the one side, and the military–technical officers on the other. However, both sides appear to have their own clearly defined career structures, and loyalties, with no evidence of any substantial switching of personnel from one to the other. The entire development process takes place in conditions of strict secrecy. Most Soviet citizens will have no knowledge of the projects until after the new systems have been deployed and examples are eventually revealed at Red Square parades or in photographs in the press.

DECISION-MAKING ON THE SS-20 – A CASE STUDY

The Soviet Union's SS-20 missile provides a good example of the complex interaction of factors which must be associated with any decision to develop and to deploy a major new weapons system.[24]

The first Soviet medium and intermediate range missiles (IRBM), the SS-4 and SS-5, were developed by the Yangel design bureau in the Ukraine during the 1950s. The SS-4, range c.2000 km, was flight-tested in 1957 and deployed from 1959. The longer-range SS-5 (c.4000 km) was deployed from 1961. At this time the USA had equivalent IRBMs, the Thor and Jupiter, deployed at sites in Britain, Italy and Turkey. The Soviet IRBMs were rather primitive, first generation liquid-fuel missiles of very low accuracy and slow reaction time, requiring up to eight hours to prepare for launch. Initially they were all installed in vulnerable, above-ground launch positions; later some were located in hardened shelters. By 1967 a peak deployment of 750 had been attained, the majority facing Western Europe, the remainder the Far East.

From about 1958 work began on the development of a solid-fuel ICBM. This was probably prompted by the US Minuteman programme and the search for a small, quicker-reaction missile with a potential for mobile basing. This new project was entrusted to a design team headed by Aleksandr Davidovich Nadiradze. Born in 1914, Nadiradze graduated from the Moscow Aviation Institute in 1940 and during the war worked as a designer on the 'Katyusha' rocket artillery system.[25] The result was the SS-13, test flown in 1965. In the same year a new medium range missile was tested, the SS-14, consisting of the top two stages of the SS-13. This appears to have been conceived as a mobile-based, second generation system, intended as a replacement for the SS-4. In 1966 Nadiradze was awarded a Lenin Prize.[26] However, solid-fuel engine technology for longer-range missiles is more demanding than liquid-fuel, posing complex problems of fuel chemistry, materials science and in-flight control. These technical difficulties were not satisfactorily overcome.

The SS-13 entered service in the late 1960s, but only sixty were deployed. The SS-14 and a similar, but longer-range missile, the SS-15, tested in 1968, failed to enter full-scale production; a few were deployed for training purposes in the Far East. Thus the Soviet Union entered the 1970s with a land-based ICBM and IRBM force consisting almost exclusively of liquid-fuel missiles: in the equivalent US arsenal the solid-fuel Minuteman predominated. At the same time, there was no viable replacement for the ageing and vulnerable SS-4s and SS-5s, the number of which was being reduced, from a peak of 750 in 1967 to around 600 in 1970.

Meanwhile the search for an effective, survivable solid-fuel ICBM with a mobile-basing potential continued, and may have been given added impetus by the 1968 trials of the multiple-warhead Minuteman III and fears that the USA might develop a missile of sufficient accuracy to pose a real threat to the Soviet silo-based ICBM force. The first test flight of the new SS-16 developed by the Nadiradze bureau took place in 1972. Once again, technological problems were encountered, the final third stage of the missile proving especially unreliable. This was probably a factor in the willingness of the Soviet side to include the SS-16 in the draft SALT II agreement: testing would cease and the weapon would not be deployed. However, before this occurred, trials had taken place in 1974–5 of a new IRBM consisting of the first two stages of the SS-16. This was the SS-20, and it was US concern that it could be rapidly converted into an ICBM that led to insistence that the SS-16 programme should be frozen.

The SS-20 has a range of up to 5000 km, three independently targetable warheads and a mobile launcher system, although it has been based at pre-prepared sites for greater accuracy. Its development enabled the Strategic Rocket Forces to undertake the long overdue replacement of the SS-4s and SS-5s. One can assume that the decision to implement this modernisation programme presented the Defence Council and Politburo with little difficulty: it had probably been planned many years before, but it was not until the mid-1970s that a viable technological solution was available. Site preparation work began in 1976; the first operational deployment in the following year. What is not clear is the initial Soviet intention with regard to the scale of SS-20 deployment and how, if at all, this intention was modified by subsequent events. By the end of 1983 the SS-5s had been withdrawn, the SS-4s reduced, and the total number of IRBMs and their aggregate warhead yield were below the levels of 1970, although the number of warheads and the accuracy of the missiles had increased. It is possible that the arms control decision to freeze the SS-16 made available additional production capacity, permitting a more rapid build-up of the SS-20 force than originally envisaged.

In 1981 Nadiradze, who is described in Soviet publications as the director and chief designer of a research institute, apparently based in Moscow, was elected a full member of the USSR Academy of Sciences.[27] On the occasion of his seventieth birthday in 1984, the Academy praised his contribution as a 'founder of a new direction in machine-building' and noted that he had trained a large number of young specialists, 'many of whom with success continue and develop the traditions of his school'.[28] The Nadiradze group may be responsible for the new, small SS-X-25 ICBM now under development, and may also be responsible for a successor to the SS-20 now under development and given the provisional NATO designation of SS-X-28. But whether these new systems will be accepted for full-scale production and deployment will depend on their technological qualities, the General Staff's assessment of military requirements, and the judgement of the highest political leadership in the light of the prevailing state of international relations and of the arms control process.

NOTES

1. *Constitution (Fundamental Law) of the Union of Soviet Socialist Republics* (Moscow: Novosti Press Agency, 1977) p. 33.

2. On these issues, see David Holloway, *The Soviet Union and the Arms Race* (New Haven and London: Yale University Press, 1983) ch. 1–3.
3. T. Wolfe, *The SALT Experience* (Cambridge, Mass.: Ballinger, 1979) p. 261.
4. *Constitution*, op. cit., p. 21.
5. On the Soviet political system, see D. Lane, *State and Politics in the USSR* (Oxford: Blackwell, 1985).
6. A. A. Epishev (ed.), *KPSS i voennoe stroitel'stvo* (Moscow: Voenizdat, 1982) p. 35.
7. D. Lane, op. cit., p. 337 (slightly modified translation).
8. V. Tolubko, *Nedelin* (Moscow: 'Molodaya Gvardiya', 1979) p. 183.
9. *Voennyi entsiklopedicheskii slovar'* (Moscow: Voenizdat, 1983) p. 769.
10. For an insight into the operation of the Politburo and Secretariat, see N. Temko, 'Soviet Insiders: How Power Flows in Moscow', in E. P. Hoffman and R. F. Laird (eds.), *The Soviet Polity in the Modern Era* (New York: Aldine, 1984) pp. 167–91.
11. *Constitution*, op. cit., p. 89.
12. See E. Jones, *The Defence Council in Soviet Leadership Decision-making*, Occasional Paper No. 188 (Washington, DC: Kennan Institute for Advanced Russian Studies, 1984).
13. *Sovetskoe administrativnoe pravo* (Moscow: 'Yuridicheskaya Literatura', 1981) p. 375.
14. *KPSS o Vooruzhennykh Silakh Sovetskogo Soyuza* (Moscow: Voenizdat, 1981) pp. 268–9.
15. A. Fedoseyev, *Design in Soviet Military R & D: the Case of Radar*, Papers on Soviet Science and Technology, no. 8 (Cambridge, Mass.: Harvard University, Russian Research Centre, 1983) p. 14.
16. On the Soviet defence industry, see D. Holloway, 'The Soviet Union', in N. Ball and M. Leitenberg (eds), *The Structure of the Defence Industry – an International Survey* (London: Croom Helm, 1983).
17. *Soviet Military Power*, 3rd edn. (Washington, DC: US Government Printing Office, 1984) p. 93.
18. See also D. Holloway, 'Innovation in the Defence Sector: Battle Tanks and ICBMs', in R. Amann and J. Cooper (eds), *Industrial Innovation in Soviet Industry* (New Haven and London: Yale University Press, 1982) and R. Berman and J. Baker, *Soviet Strategic Forces: Requirements and Responses* (Washington, DC: The Brookings Institution, 1982) Appendix A.
19. *Soviet Military Power*, op. cit., p. 97.
20. Korolev died in 1966, Yangel in 1971 and Chelomei in 1984; the Yangel OKB now appears to be headed by V. F. Utkin, the other two heads are not known to the author,
21. *Sovetskaya Voennaya Entsiklopediya*, vol. 2 (Moscow: Voenizdat, 1976) p. 513.
22. J. Kraft in *The New Yorker*, 29 July 1979 cited by D. Holloway, 'Military Power and Political Purpose in Soviet Policy', *Daedalus* (1980) p. 30.
23. This section draws on a number of Soviet and Western works, including V. Tikhomirov, *Organizatsiya, planirovanie i upravlenie proizvodstvom letatel'nykh apparatov* (Moscow: 'Mashinostroenie', 1978); I. Kuskin and S. Moiseev (eds), *Organizatsiya, planirovanie i upravlenie proizvodstvom*

radioelektronnoi apparatury (Moscow: 'Mashinostroenie', 1979); I. Fakhrutdinov, *Raketnye dvigateli tverdogo topliva* (Moscow: 'Mashinostroenie', 1981) and D. Holloway, 'Innovation in the Defence Sector' in R. Amann and J. Cooper (eds), op. cit.

24. Unless otherwise indicated, this case study draws on the following, Western works: R. Berman and J. Baker, op. cit.; R. Garthoff, 'The Soviet SS-20 Decision', *Survival*, 25 (May–June 1983); T. Wolfe, op. cit.
25. *Bol'shaya Sovetskaya Entsiklopediya, Ezhegodnik* (Moscow: Izd. 'Sovetskaya Entsiklopediya', 1982), p. 579; V. Shvedov, *KPSS – organizator voennogo proizvodstva v 1941–1942 gg.* (Leningrad: Izd. Leningradskogo Universiteta, 1982), p. 113.
26. *Bol'shaya Sovetskaya Entsiklopediya, Ezhegodnik*, op. cit., p. 579.
27. *Vestnik Akademii Nauk SSSR*, (1982, no. 4) p. 129; *XXVI s"ezd Kommunisticheskoi Partii Sovetskogo Soyuza, Stenograficheskii otchet*, vol. 3 (Moscow: Politizdat, 1981) p. 428.
28. *Vestnik Akademii Nauk SSSR* (1984, no. 10) p. 132.

2 The United States of America

The US was the first nation to produce a nuclear weapon. Since 1945 a vast and extraordinarily powerful organisation has developed for designing, developing and producing thousands of warheads and delivery systems each year.

The decision-making bodies which encourage, permit or restrain this process are of intimidating complexity. Four intelligence agencies are involved, key sectors of the Department of Energy and the Department of Defence, the Air Force, Army and Navy, giant industrial corporations, the White House and the Office of Management and Budget, the National Security Council, the State Department, advisory councils, think tanks, and congressional committees.

The complexity is increased because, in America particularly, the relative influence of these various bodies depends on who occupies the key positions in them. Some personalities exert strong influence although they have no significant status, while other high positions lose significance when the incumbent is disinterested in nuclear issues.

There are in addition a number of powerful themes constantly affecting the process – accelerating it, sometimes restraining it. These themes – threat assessment, technological possibility, bureaucratic momentum, interservice rivalry, corporate pressure, foreign policy, budgetary and presidential influence – will crop up one by one through the descriptions that follow.

We shall examine in turn each of the major bodies involved in decision-making on nuclear weapons, beginning with those who provide the initial briefings on which weapons needs are based, and working through the process of weapons development decisions to the role of Congress.

THE INTELLIGENCE COMMUNITY

The agencies, departments and boards which make up the intelligence community are under the jurisdiction of the executive branch with the President at the head.

These organisations are part of the nuclear weapons decision-making process in fundamental ways: they produce the annual 'threat assessment' upon which the Department of Defence (DoD) bases its annual projection of military needs; they are an integral part of the satellite surveillance system critical to targeting of weapons; they set up and maintain military communications systems for the National Command Structure.

The US intelligence community, which employs upwards of 100 000 people worldwide, has as its cornerstone the Central Intelligence Agency (CIA). Within the Department of Defence are several intelligence bodies: the National Security Agency (NSA), the Defence Intelligence Agency (DIA) and the individual intelligence units of the services.[1] The State Department has the Intelligence and Research Bureau. Several other executive departments, such as Energy, and Commerce, maintain foreign intelligence units.

The key area of intelligence work on nuclear weapons produces the annual 'threat assessment', an overview of levels of power of other nations, their nuclear research and development projects, budget allocations to new systems, production testing and deployment plans. The Pentagon uses this as the basis for formulating weapons 'needs analysis'. There are some who maintain that the needs analysis is a serious response to the threat assessment, which is in turn a realistic appraisal of potential risks and dangers. Others consider the assessment an annual indulgence in portraying the Russian adversary as so alarmingly treacherous that an all-out expansion of capacity and R & D is necessary to protect the United States. It does however become the baseline, and in its building of an adversarial picture the intelligence community sets the stage for nuclear weapons decision-making.

To further fuel the fire, in what is widely viewed as a public relations exercise, the Pentagon publishes annually a glossy 140 page edition of *Soviet Military Power* – with coloured maps and charts and striking photographs. The 1985 edition claimed test firings of fifth generation ICBMs.

Figure 2.1,[2] which follows, illustrates how the National Intelligence Estimates are produced each year, starting at the focal point of the Director of Central Intelligence. He assigns tasks to the four main agencies described below, who send back drafts for the National Intelligence Estimate. It is these Estimates which form the base for the crucial threat assessment, which goes to the President and the National

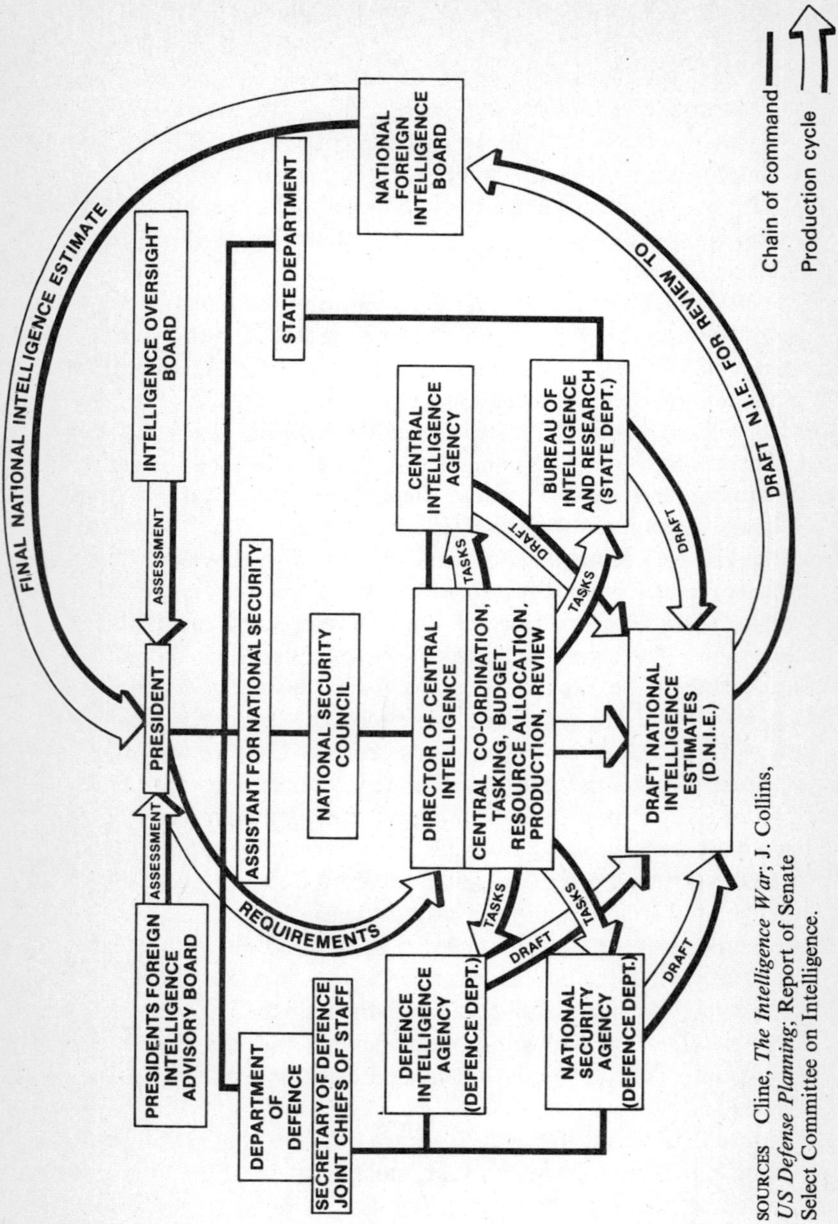

SOURCES Cline, *The Intelligence War*; J. Collins, *US Defense Planning*; Report of Senate Select Committee on Intelligence.

FIGURE 2.1 *The US defence intelligence cycle*

Security Council (NSC) as well as to the Pentagon and State Department.

The Central Intelligence Agency (CIA)

Chartered as part of the National Security Act of 1947, the CIA serves technically under the National Security Council, the President's inner cabinet on national security. The CIA is the central co-ordination point for tasking, budget, resource allocation, priorities production and review for all intelligence work.

The Director of Central Intelligence (DCI) serves in the Cabinet, is appointed by the President and serves at his discretion. He is confirmed by the Senate. He is principal advisor to the President and the NSC on foreign intelligence, serves as director of the CIA, and it is he who sets the objectives and policies for the entire intelligence community.

In terms of accountability 'the intelligence community operated almost forever without having to report to anyone' according to Congressman C. W. Bill Young, a Florida Republican and member of the House of Representatives Permanent Select Committee on Intelligence,[3] but members of the committee say they believe the agency is no longer the 'rogue elephant' that it was in the 1960s and 1970s.

Secrecy being the keynote of the agency's work, it comes as something of a surprise when its main entrance in McLean, Virginia, is marked by an immense, overhead green and white road sign off Virginia Route 123, with an arrow, CIA.

The Defence Intelligence Agency (DIA)

Created in 1961 through amendments to the National Security Act, the DIA is an agency of the Department of Defence, and is directly responsible to the Secretary for Defence. Its mission is to satisfy the foreign military intelligence requirements of the Secretary of Defence, Joint Chiefs of Staff and major Department of Defence (DoD) divisions. The Director, a uniformed officer of flag rank, is appointed by the Secretary for Defence and reports to the Under-Secretaries for Research and Engineering and Policy and to the Chair of the Joint Chiefs of Staff.

DIA does not collect raw data except in its capacity of operating the Defence Attache System of diplomatically accredited military observers in overseas embassies and stations. Instead, it collects, manages, co-ordinates, controls and analyses data from other DoD divisions and from the CIA and NSA with which it co-operates.

The National Security Agency (NSA)

The National Security Agency operates from Fort Meade, Maryland and is responsible for the direction and performance of highly specialised technical functions to protect US communications and produce foreign intelligence information. Established by Presidential Directive in 1952, it is the electronic headquarters for traffic analysis and surveillance (monitoring the Soviet ICBM test sites and staffing surveillance aircraft such as the one involved in the Korean air lines incident in 1983). In this capacity its work is important to issues of treaty compliance and verification. Moreover, the NSA supervises the necessary closed communications for military command and control and maintains the National Command Structure in nuclear war planning. The Agency Director is appointed by the Secretary of Defence. Although it is within the DoD, NSA has a national intelligence role extending beyond the armed forces into the entire government, gathering intelligence for national policy-makers.

The Bureau of Intelligence and Research

This Bureau, which is at the State Department, co-ordinates foreign policy intelligence gathering, research and analysis. It provides information on the political responses of other nations to US foreign policies; for example, on European reaction to cruise and Pershing missile deployment. Through its Office of External Research, the Bureau maintains liaison with other institutions engaged in foreign affairs research, including technology exchange information for arms control.

The President's Foreign Intelligence Advisory Board (FIAB)

Created by President Reagan in October 1981 to improve the quality and effectiveness of intelligence available to the US, the Board reports to the President twice a year. Members of the Board, the Chair and Vice-Chair are appointed by the President, serve at his pleasure, and have free access to materials they need.

The National Foreign Intelligence Board (NFIB)

Made up of members and observers from all the major military and civilian components of the intelligence community, the NFIB advises the Director of Central Intelligence on the draft National Intelligence Estimates before they go to the President each year.

The data collected is differentiated most by the way each agency interprets it: political and ideological shading of interpretation, while inevitable,[4] can be pivotal in setting the direction of an administration's policies.

Intelligence monies are classified data, but joint budgets of the agencies were estimated at $15 500 000 000 for financial year 1984. (The US financial year starts in October, thus the year starting October 1983 will be written FY 1984.) Under laws governing the CIA, Congress is supposed to be kept fully informed of intelligence activities. Congress has no approval power over specific agency operations, but the congressional oversight process (select committees in both houses) has had some success in blocking actions (those of which Congress is aware) by going directly to the President. Congress also controls the agencies' purse strings and as a last resort can use this power to force changes.[5]

THE DEPARTMENT OF ENERGY

Nuclear weapons have always been at the heart of what is now the Department of Energy (DoE), going back to President Roosevelt's approval in June 1942 of the development of the atomic bomb. The Manhattan Project produced the first successful detonation of a nuclear device in July 1945, and the bombs dropped on Hiroshima and Nagasaki.

The present Department of Energy, created in 1977, represents years of effort to combine the resources needed to administer nuclear weapons, as well as overall US energy programmes. The weapons programme takes a growing share of the DoE budget, from 25% in 1980 to 63% in FY 1985.[6] The dominant theme, reiterated throughout the years, is that the decisions and physical preparations regarding nuclear warheads remain in civilian control. Some feel that the current DoE configuration honours this intent. Others feel that it only provides an easily administered programme for implementing exactly what the services want, plus an emphasis on weapons' technological advancement. And still others feel that DoE is so riddled with structural flaws that allow the nuclear weapons laboratories and nuclear energy industry the opportunity to use official channels for advancing their own agendas and interests.

Structurally, development and production of nuclear weapons systems are the combined responsibility of the DoE and the DoD. The design, development, testing and production of the actual warheads is concentrated in the DoE. At the time of writing, a presidential commission is examining the advisability of merging DoE nuclear weapons activities into DoD to reduce warhead costs. The DoD is responsible for the military characteristics and development priority of nuclear weapons; for custody and maintenance of the stockpile – for design, development testing and production of delivery systems; for training and deployment of forces for their use. In conjunction with the Intelligence Community, DoD establishes the threat assessment on which nuclear weapon need requests are based.

The Military Liaison Committee (MLC) is the key link between the DoD and the DoE (except for direct communications between Secretaries). In view of the concern over civilian control of nuclear weapons,[7] it is somewhat surprising to find that it is chaired by a military officer of flag rank who serves concurrently as Assistant Secretary of Defence for Atomic Energy. The Committee's offices are in the Pentagon and the staff is primarily active duty military personnel.

DoD has the authority to make recommendations to the DoE on issues relating to military applications of atomic energy as it deems appropriate. This is particularly significant now in areas such as the Strategic Defence Initiative programme, for which DoE is developing the X-ray laser pumped by a nuclear explosion.

Within the DoE the weapons programme is the responsibility of the Assistant Secretary of Energy for Defence Programmes, who has four Deputies: for Military Application, Nuclear Materials, Security Affairs and Intelligence. Under the Office of Military Application fall the three famous nuclear weapons laboratories, and operation of facilities for the production and testing of nuclear warheads.

Los Alamos and Lawrence Livermore

Every nuclear warhead in the United States was conceived and designed either at the Los Alamos National Scientific Laboratory, New Mexico, or the Lawrence Livermore National Laboratory in the

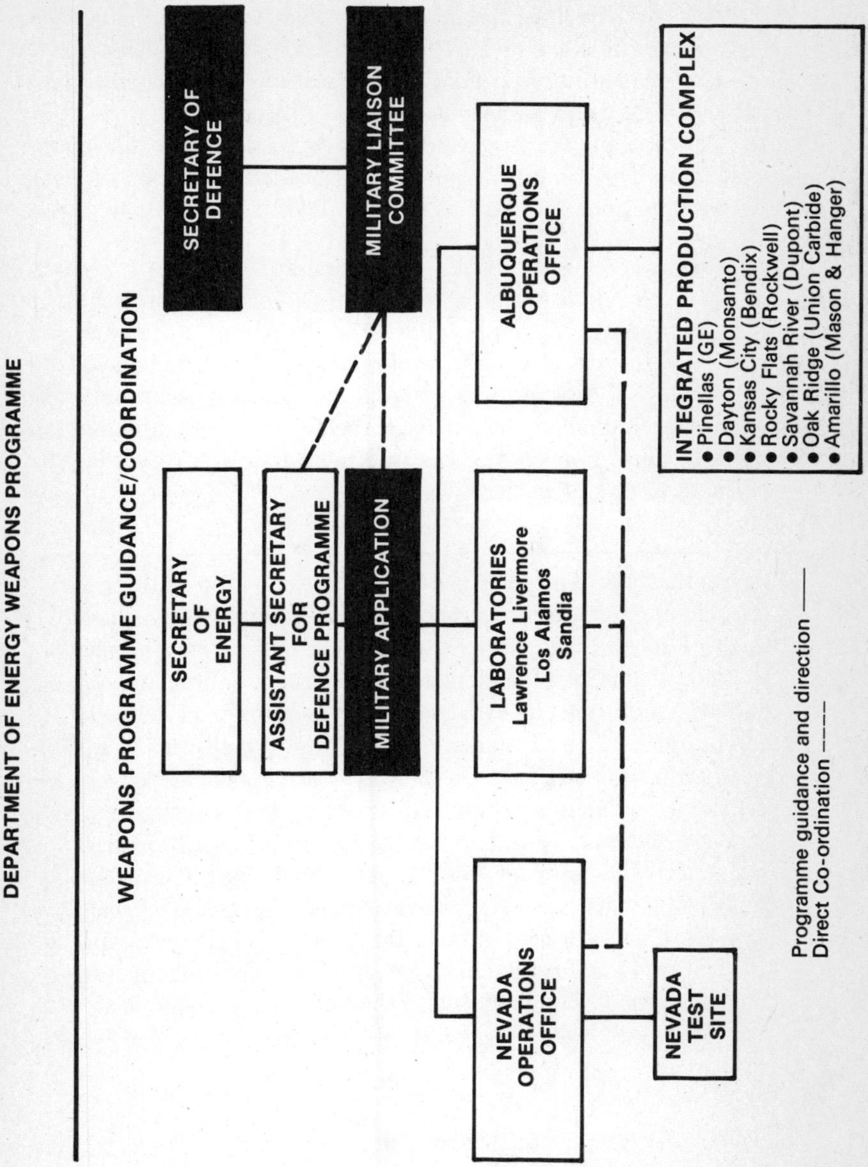

FIGURE 2.2 *Department of Energy weapons programme*

Livermore Valley, an hour from downtown San Francisco. Officially, they are operated by the University of California, which lends its name, prestige and academic respectability to their work, but control and funding are in the hands of the Department of Energy.

About 7000 people are employed at Livermore and almost as many at Los Alamos. The combined annual budget of the two laboratories is close to $900 million, with half to two-thirds of it related to weapons work.

As competitors (Livermore was established in 1952 to spur research on the hydrogen bomb, which was thought to be lagging at the older Los Alamos laboratory), the two have added greatly to the pace and scope of US nuclear weapons development. They have played an important role in lobbying for new weapons systems and against arms limitation treaties that might cut back their work. They are credited with a major part in preventing US agreement to a treaty banning the underground testing of nuclear weapons.

> 'From the days of his 1976 campaign, Carter endorsed the idea of even more restrictive limits on nuclear testing. His overtures to the United Nations in October 1977 met with a positive response from Soviet leaders and formal negotiations opened later that month in Geneva. In the early months of 1978, it was quite clear that the President was willing to pull out all the stops for a total test ban. On 20 May, Carter signed Presidential Decision Memorandum 38 instructing US negotiators to pursue a five-year total test ban treaty. In August, however, both Kerr (Director of Los Alamos) and May (Associate Director of Livermore) appeared before the House Armed Services Committee to attack the five-year testing hiatus. Laboratory Directors Batzel and Agnew sent critical letters to key members of Congress.'[8] There was no five-year test ban.

Through the years, the laboratories have been a training and recruiting ground for leadership positions in the US defence establishment. A former Livermore Director, Harold Brown, served as President Jimmy Carter's Secretary of Defence. Former Los Alamos Director Harold Agnew, now a member of the White House Science Council, testified in April 1973 before the House Armed Services Committee:

'I know that we at Los Alamos have a small but very elite group that meets with outside people in the defence community and in the various think tanks. They are working very aggressively, trying to influence the DoD to consider using these neutron weapons which would be very decisive on a battlefield yet would limit collateral damage that is usually associated with nuclear weapons.'

Sandia National Laboratories

Sandia is one of the US government's largest research and engineering facilities, employing over 8000 people at its headquarters at Kirtland Air Force Base, Albuquerque, New Mexico, a smaller laboratory at Livermore, California and a test range near Tonopah in Nevada.

Sandia is operated for the government by a subsidiary of the American Telephone and Telegraph Company. The three laboratories, and the plants which manufacture the warheads, are known as GOCO – government-owned-contract-operated.

Sandia alone among the laboratories has the responsibility for developing the non-nuclear components of nuclear warheads. Although its original mission was weapons engineering development and bomb assembly for the Manhattan Project at Los Alamos, Sandia does no actual manufacturing or assembling of weapons today; this work is performed by other government-owned facilities, using design information provided by Sandia and the other weapons laboratories.

One of Sandia's specialities is ensuring a weapon's resistance to radiation from other nuclear explosions. As soon as President Reagan had launched the Star Wars concept on 23 March 1983, three prominent Sandian scientists were directing studies on the new initiative as part of the Defensive Technologies Study Team.[9] Sandia has since forged ahead with plans for a $70 million testing facility.

Warhead Manufacture

A number of well-known major industrial corporations manufacture nuclear warheads under contract to DoE in government-owned facilities. Put briefly, the process involves first the enrichment of uranium in the isotope uranium-235 in gaseous diffusion plants operated by Martin Marietta Corporation and Goodyear Atomic Corporation. This fuels the reactors that make fissionable plutonium at Savannah

River in South Carolina, the country's major nuclear materials production centre, which employs 8000 people and is operated by Du Pont, and at the Hanford Reservation, a 570-square-mile desert tract in Southeastern Washington State, run by United Nuclear Industries and Rockwell International. Most of the uranium-235 and lithium parts for US nuclear weapons are fabricated in a 500-acre bomb factory known as the Y-12 Plant, part of the huge Oak Ridge complex in the foothills of the Great Smokey Mountains in Eastern Tennessee. This plant has been run by Union Carbide for the DoE since 1947.

The plutonium trigger systems that ignite thermonuclear bombs are made at the Rocky Flats Plant, sixteen miles upwind from Denver, Colorado, run by Rockwell International. Detonators and timers are produced by Monsanto at the Mound laboratory in Ohio, while neutron generators are made by General Electric at the Pinellas Plant in Florida. Most of the non-nuclear parts – guidance systems, locking devices and environmental sensing components, – are made in Kansas City by the Bendix Corporation.

There are more than 2000 separate parts in a typical nuclear warhead. They all come together at a final assembly plant in the Texas panhandle, the Pantex Plant, operated by a private company, Mason and Hanger, Silas Mason Co. The plant also makes chemical high explosive charges, disassembles retired warheads and bombs, and modifies warheads already in the stockpile.

At Pantex, the warheads are loaded into trucks for shipment by interstate highway to Army, Navy and Air Force bases which deploy nuclear weapons in the continental United States. Thousands are shipped overseas to the Far East and Europe.

It will be clear even from this brief overview that the nuclear weapons complex managed by the DoE is a major nationwide enterprise, with 52 500 employees and $25.4 billion in assets.[10] If nuclear weapons were traded on the US stock exchange, DoE would rank ninth in the Fortune 500.[11] The only comparable enterprise in the world is the Ministry of Medium Machinebuilding in the Soviet Union, responsible for all stages of the production of nuclear warheads, as well as playing a major role in the civilian nuclear power programme.

THE DEPARTMENT OF DEFENCE

Since the days of the Revolutionary Militia, American public policy opposed a standing army. But the experience of World War II provided

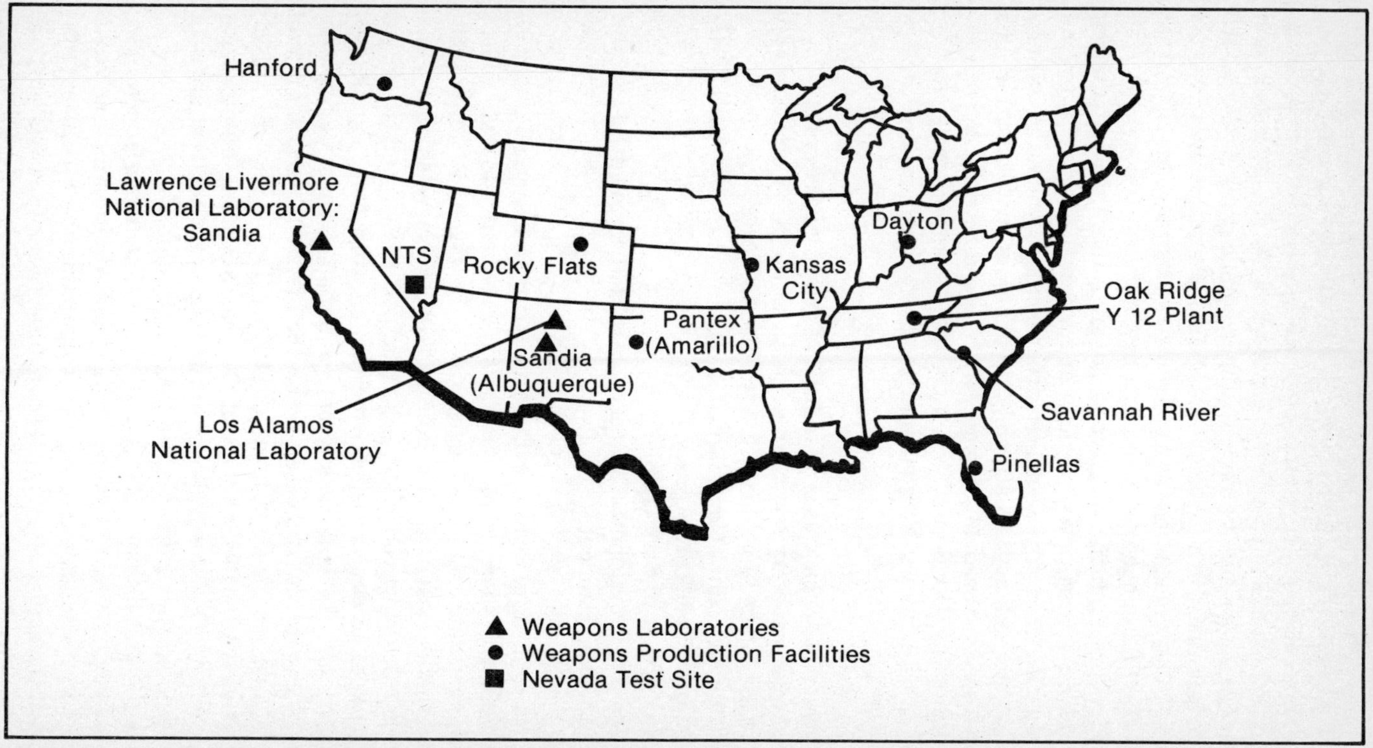

FIGURE 2.3 *Weapons laboratories and production facilities.*

a consensus to build a professional, unified military capability able to defend national security. The public's will was to prevent the vulnerability experienced at Pearl Harbour and preserve the military superiority inherent in possession of atomic weapons. Public opinion polls of the time reveal that Hiroshima and Nagasaki were seen more as the ultimate triumph of American know-how and moral rightness than as harbingers of a dangerous future or a new military strategy.

In President Truman's message to Congress after the war, he stated that 'there is enough evidence now at hand to demonstrate beyond question the need for a unified department, beyond the existing decentralised military departments'. After two years of continued studies and congressional hearings The National Security Act of 1947 was passed by Congress. That Act and its amendments are the framework for today's overall national security configuration. The Act moved to establish:

A Secretary of Defence as civilian head of the Armed Services
A third major military service in addition to the Navy and Army by separating the Army Air Corps into the Air Force
The National Security Council (NSC)
The Central Intelligence Agency (CIA)
The Joint Chiefs of Staff (JCS)
Research and Development as a key part of the growth of the new military establishment.

In less than forty years the Department of Defence has grown to include three million people – one million civilians and two million military personnel. Its headquarters, the Pentagon, is a five-sided granite building just across the Potomac River from Washington, DC. Its tree-shaded inner courtyard, helicopter pad and sixty seven acres of parking lots are clearly visible from the air when flying into Washington airport. The building's seventeen-and-a-half miles of corridors form five concentric 'rings' and house the offices of the military and civilian leaders of the Department of Defence (DoD). The 'E' or inside ring is reserved for the Secretary of Defence (SECDEF) and his most senior staff. DoD offices spill over to buildings in the adjacent towns of Arlington and Alexandria and sites across the globe.

Neither warheads nor delivery systems are manufactured here. This is the policy, planning, strategy, budget and decision-making headquarters for America's defence. As we examine the process of procurement, the actual ordering and commissioning of nuclear weapons systems, we shall come across the agencies, departments, councils and committees which play a key part in the process. They are identified by an asterisk on the (simplified) organisational chart of the Pentagon (see Figure 2.4).

A useful way of understanding the complex process of procurement is to look first at how one weapon is produced – the actual acquisition – and then to examine the annual processes which determine *which* and *how many* weapons are produced. Nuclear weapons systems follow a dual procedure at every stage of their development, that of the warhead and that of the delivery system – (the missile or launcher which takes the warhead to its destination) – and procurement is no exception. Warhead acquisition is discussed first.

Warhead Acquisition

The Department of Energy, as we have already seen, designs, tests and builds the warheads, and receives funds to do so through the congressional authorisation and appropriation system. But decisions on the development phases are largely the preserve of the DoD.

Phase 1: Weapon Concept

This is essentially research and may concern the modification of an existing weapon or the development of a new one.

Nuclear weapons are differentiated by more than just their megatonnage. A range of factors interplay in varying their size, weight, shape and destructive capability. For example, depending upon the accuracy of a delivery vehicle (by land, sea or air) and the nature of the target, DoD strategists request a weapon with a certain destructive capability to accomplish a specific task. The combinations of nuclear weapons effects, blast, heat and radiation is not infinite. But scientists and engineers can design factors up or down, as demonstrated in the so-called neutron bomb, whose enhanced radiation and reduced blast and heat inflict maximum human damage with a minimum of property dislocation.

Phase 2: Feasibility Study

Before this can be done a 'request for engineering and development' has

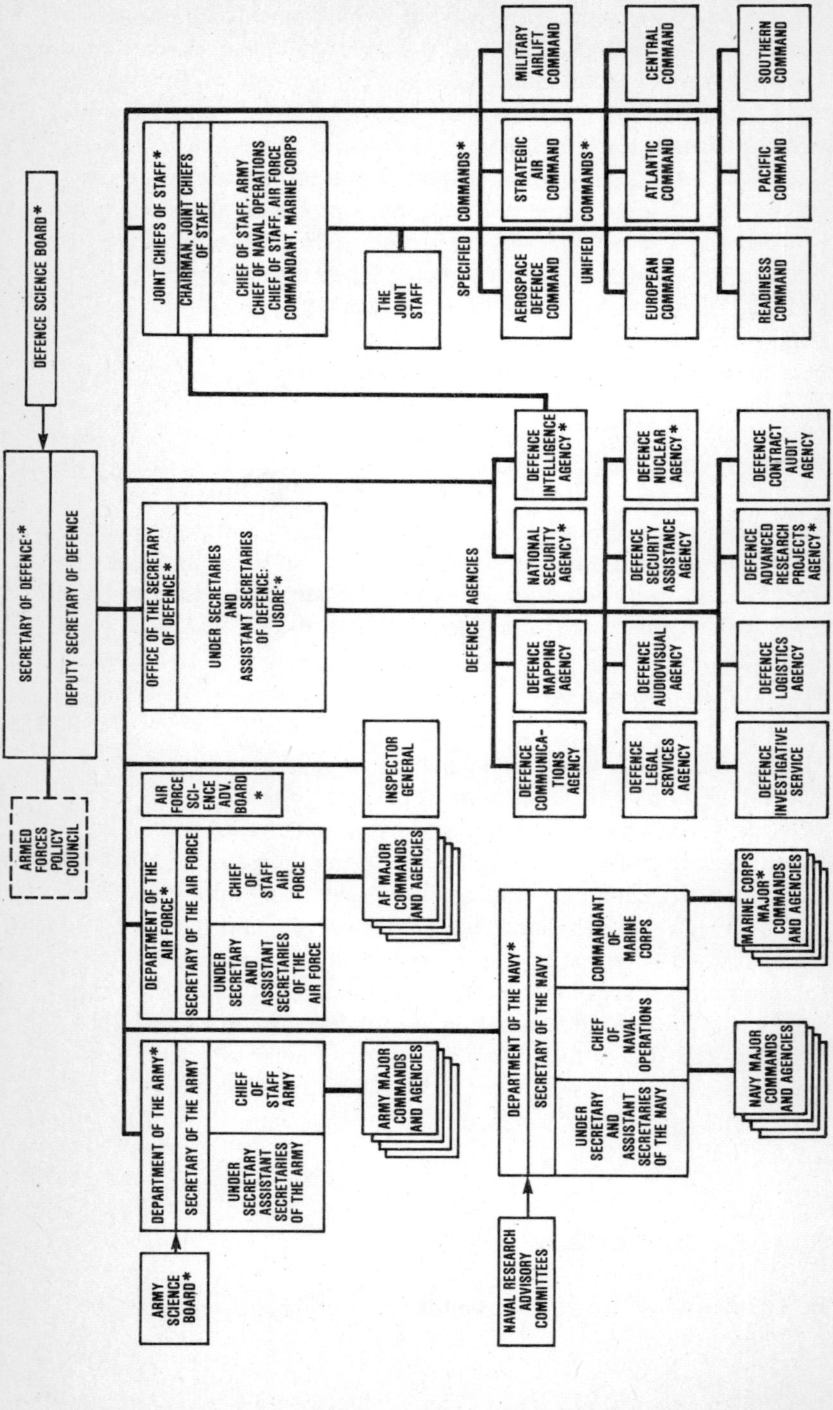

FIGURE 2.4 *Department of Defence*

to be submitted to the Under-Secretary of Defence for Research and Engineering (USDRE)* which must outline the military characteristics of the weapon. This DoD request, and DoE acceptance of it, constitutes in effect a contract to develop and prepare for production of a warhead. When approved, USDRE has authority to designate the 'cognizant' military department, probably the one which submitted the request, to chair a Joint Feasibility Study, with DoE participation. The three main military departments and the Joint Chiefs of Staff* review Phase 1 and Phase 2 studies annually.

Phase 3: Weapon Development

If the evaluation is favourable, the Secretary of Defence will authorise a Phase 3 development programme, the formality for which is a Phase 3 development request to the DoE through the Military Liaison Committee.

From this point the cognizant military department will assign a lead project officer, and conclude an agreement with DoE on the division of responsibilities. Here other military departments which are interested in the project may designate a department spokesperson for participation. The Defence Nuclear Agency (DNA)* may assign a non-voting project officer to the development project to observe and to provide technical assistance and support. Stockpile-to-target sequence is a document to be completed within 60 days of the Phase 3 letter initiating the project. Preliminary, interim and final reports in warhead development are reviewed by the Design Review and Acceptance Group (DRAAG) with representatives from Army, Navy and Air Force presided over by the cognizant department chair. The DRAAG report is the vehicle by which the MLC notifies the DoE that the design is acceptable and production of the warhead may begin.

When production is complete, the warhead and delivery system are assembled and maintained as live weapons at the designated reception areas. This is the case of systems designated for immediate nuclear capability, such as Poseidon, Polaris and Trident submarines, B-52 bombers, Minuteman Missiles and Air Launched Cruise Missiles. Other weapons for dual capacity systems, conventional and nuclear, are stockpiled on site under DNA supervision. They will travel parallel with the delivery system, such as weapons to go on board an aircraft carrier, but will not be mated with the delivery vehicle until the order is received to go nuclear.

DNA oversees a very detailed system for overall supervision, allocation, storage and verification of nuclear weapons. Each weapon is identified by a serial number and tracked by that number. All details are stored in the Worldwide Military Command and Control System. Information transmission and accountability is allegedly strict and in theory DNA should know at any time where each weapon is, where it has been, and who has handled it.

Warhead Annual Planning Cycle

1. The Joint Chiefs of Staff submit biennially to the Secretary of Defence the proposed Nuclear Weapon Development Guidance. This is based on the requirements of the services – the Army, Navy, Air Force and Marines – to satisfy projected operational needs, over the next ten to fifteen years and on the priorities of the Unified and Specified Commands for particular warhead qualities and technological goals.
2. The Under-Secretary of Defence for Research and Engineering, in co-ordination with other assistants to the SECDEF who have nuclear weapons responsibilities, prepares and transmits to the DoE the Annual Nuclear Weapons Development Guidance letter, through the Military Liaison Committee. This tells the DoE what the DoD is likely to want in new warheads, gives development priorities, and identifies technological goals for future weapons.
3. The Nuclear Weapons Stockpile Memorandum, an ongoing analysis and projection of stockpile needs determined by the above, is prepared by the Defence Nuclear Agency and presented to the President through the National Security Council.
4. Presidential approval is given, authorising the stockpile for five years and granting long lead procurement authority for an additional ten years. All of these documents are classified.

The JCS submits a Nuclear Weapons Deployment Plan each year to the Office of the Secretary for Defence, who co-ordinates it with the State Department, and then submits it to the President. Presidential authorisation then permits weapons to be deployed. The plan provides authorisation to deploy weapons overseas in time of peace and war and provides a breakdown of warheads designated for allied military use under nuclear Programmes of Co-operation (see page 218).

Major Delivery Systems Acquisition

Every two weeks, on average, six men meet in a conference room on the first floor of the Pentagon and decide to spend hundreds of millions of dollars on new weapons systems. This group is known as the Defence Systems Acquisition Review Council (DSARC). Its members include the Under-Secretary of Defence for Research and Engineering (USDRE), the Under-Secretary of Defence for Policy, the DoD Comptroller, the Director of Programme Analysis and Evaluation, the Chairman of the Joint Chiefs of Staff and the Assistant Secretary of Defence for Manpower and Logistics.

No major R & D programme is undertaken (those costing more than $200 million in R & D are 'major'), and no billion dollar production contracts are signed without the approval of this Council. It supervises the work of the individual services, who are primarily responsible for the ordering and production of their 'own' weapons. As soon as a new weapon is decided upon, a project office for that weapon is set up within the service concerned. For example, an office to direct the development and production of Minuteman, called the *Minuteman System Project Office*, existed for many years within the Air Force department.

This system is distinctly different from that in Britain where the Procurement Executive handles all weapons production. It has come in for severe criticism on two main grounds. Firstly, military organisation in the US is dominated by rivalries between the services, and this is one underlying cause of bloated budgets.[12] Service 'fiefdoms' have been allowed to flourish. This leads to duplication of weapons systems on a staggering scale. Secondly, because production of a weapon is supervised by the same people who placed the original order, this inhibits the cancellation of doubtful programmes which fail to meet expectations, and leads to the widespread practice of re-designing weapons after production has begun.

It takes an average of about ten years before a weapon, once conceived, is ready for production. The DSARC reviews embryonic weapons programmes at critical stages in their development, known as 'milestones'.

Milestone One comes at the end of the process called *Mission Need Determination* when the particular service is exploring means of addressing a military problem, theoretically with no specific weapon in mind. In practice, however, each service receives powerful blasts of new technological possibilities from a number of sources: from within its own laboratories, through periodic meetings with scientists working in other national laboratories and from the research undertaken by giant industrial corporations.

Two other bodies of key importance must be mentioned here. The *Defence Advanced Research Projects Agency* (DARPA)* has the 'responsibility to manage high-risk, high pay-off basic research . . . to carry advanced programmes to feasibility demonstration and then transfer them to the appropriate military service'.[13] Although it is a small agency, DARPA is highly influential and has overseen the development of a number of new technologies including 'stealth' radar evasion technology.

The *Defence Science Board* (DSB)* and the science boards which advise each service also play a key part. Their members are the scientists who invent weapons and technologies; the engineers who make the pieces work; the retired military who ordered and directed the systems and are now in high corporate posts. There is an array of 'formers', civilians who held high Pentagon posts in previous administrations, corporate officers who represent the success stories in management and product development, university presidents, laboratory directors and a smattering of former elected officials. All of them have worked their way up the weapons ladder to become the experts who are called upon to 'provide independent advice and information to military departments' and to the SECDEF.

Equipped with updated security clearances, staffed by Pentagon employees, briefed by military officers and other selected participants, Board members are called upon to 'provide broad policy advice and specific technical recommendations for solving particular problems. The committees conduct their work primarily through the *ad hoc* panels and task forces which perform specific tasks or undertake specific studies'.

But spiralling defence costs and failed weapons have given rise to questions about the Board's usefulness as 'independent' advisors. Public and private questions prompted an investigation of the Service Advisory Boards by the Government Accounting Office. At the same time the Defence Science Board Chair requested the Inspector General of the DoD to launch a similar investigation of DSB. Both reported in September 1983.

In summary, they found a systematic lack of adherence to legal procedures, of 'balance' in appointments to the Boards and panels, a non-recording of potential conflicts of interest which made a determination of fact impossible and an approximate 15% overlap of participation.

Milestone Two was formerly the point at which DSARC reviewed the idea for a new weapon before it went into full-scale development. Under new acquisition guidelines, however, this review point has been allowed to lapse. Entry into full scale development is a crucial point in the life of a weapons system: while it may already be consuming tens of millions of dollars per year, it has not yet begun to cost enough to warrant close Congressional Committee attention. This is the time when a weapon acquires its bureaucratic and contractor 'constituency' which makes later cancellation virtually impossible. The Secretary of Defence must approve the passage of a weapon into the next phase during which the services and selected contractors design, build and test prototypes. When this phase is complete, the weapon is ready for mass production. On paper, there is a DSARC milestone that the weapon must pass before it can be produced and deployed, but under the Reagan administration, the services do not ordinarily need DSARC permission to proceed.[14]

DSARC's performance has been persistently criticised by the Pentagon's own Inspector General. Since 1982 audits of the committee's decisions in sixteen different weapons programmes have showed that DSARC members routinely ignore DoD procurement regulations, rarely scrutinise a weapon during early stages of its development, fail to demand detailed information on weapons' defects and rarely demand an explanation for delays. 'The DSARC process, as presently formulated, does not provide sufficient insight to the decision-makers.'[15] Richard Delauer, who chaired the DSARC until his resignation as Under-Secretary of Defence for Research and Engineering, sharply disagreed with these assessments, saying that DSARC was functioning as recommended.

The office now called Programme Analysis and Evaluation (once known as Systems Analysis) was set up by MacNamara in the early 1960s precisely to provide critical independent analysis of all aspects of weapons programmes. Its Director sits on DSARC, but its influence has waned since MacNamara's days. With rare exceptions DSARC members defer to 'military judgement' or the confidence of the technologists that a new system is needed and will perform.

> One analyst, a former voting member of DSARC, suggests that such imprudent deference is driven by two factors: firstly, the Secretary of Defence does not support challenges to the prevailing service wisdom; secondly, no-one is held accountable when costs overrun or systems just don't work. Programme managers have moved on, senior service people have been promoted or retired, civilian leaders have left. Even when the responsible people are still in their jobs, no-one makes them take responsibility.[16]

Some operative changes were made in January 1985 which will reallocate responsibility for weapons purchases. The Under-Secretary of Defence for Research and Engineering formerly controlled most aspects of both development *and* acquisition of new weapons – in effect, enabling the Pentagon's top scientist to pass judgment on the worth of his own creations. Primary responsibility for overall production policy was handed in 1985 to a new office. The effect has been to drive a wedge between those responsible for R & D and those responsible for production. But the actual production and acquisition of each new weapons system remains firmly under the control of the individual services. A group of senior corporate executives, former members of the President's Private Sector Survey on Cost Control (a high-level panel handpicked by the White House to identify waste and inefficiency throughout the government) has described DoD weapons buying as costly, chaotic, inefficient and undisciplined, and in need of drastic structural change[17] (see page 73). Again, some changes have been made. The US Navy, in a major policy shift which reflected its difficulties with defence contractors, decided in March 1985 that 40% of officers promoted to admiral must have specialised in weapons procurement.

Having examined warhead acquisition and the warhead annual planning cycle (which determines which and how many warheads are produced), and the acquisition process for major delivery systems, it now remains for us to dissect the annual planning cycle for delivery systems, known as PPBS.

Planning, Programming and Budgeting System (PPBS)

This is the central feature of US defence budget-making, and forms the framework for decisions on how many weapons systems are produced, and of what kind. Its essence is explained in Figure 2.5.

It starts with the *Threat Assessment* ascertaining the threat to US national security. The Secretary of Defence uses this as the basis for his annual *Defence Guidance* – his official statement outlining the strategy the DoD should pursue to meet that threat: due every January, the Defence Guidance tells each military service how much money it should plan to spend to achieve its part in those goals over the next five years. Development of the Defence Guidance is probably the most important step in the entire budget process, because it sets distinct goals and establishes priorities for achieving them.

The Army, Navy and Air Force respond by making concrete proposals to achieve their assigned goals, called *Programme Objectives Memoranda* (POMs). These five-year blueprints for force planning set out how many weapons and of what kind each service wants. This is where new weapons start; weapon modernisation, research, development, testing and procurement are introduced for funding. Throughout the year, consultants, contractors and corporations lobby for inclusion of their programmes in the services POM. Similar pressures come from within the services' own weapons and research laboratories. DoD personnel responsible for ongoing 'pure' research funded by DoD in universities, corporations and government must assess those programmes for continuation or for 'weaponising'. They in turn are lobbied by scientists for programme continuations.

The service POMs are submitted at the end of July. All projects with one year price tags exceeding $25 million are reviewed by the Defence Resources Board (DRB) – the focal point of PPBS. Meeting at least monthly, the Board makes most of the decisions that are made during the defence budget process. The Chairman of the JCS is the only military member of the Board, which otherwise consists of the top fourteen DoD civilians and a representative of the Office of Management and Budget. It is chaired by the Deputy Secretary of Defence, and its rising importance (it had only six members when created in 1979) has come partially at the expense of the systems analysis office, mentioned earlier. The DRB also considers the evaluation by the Joint Chiefs of each services POM – that report, called the *Joint Programme Assessment Memorandum*, judges whether the POMs, taken as a whole, satisfy the mandate given.

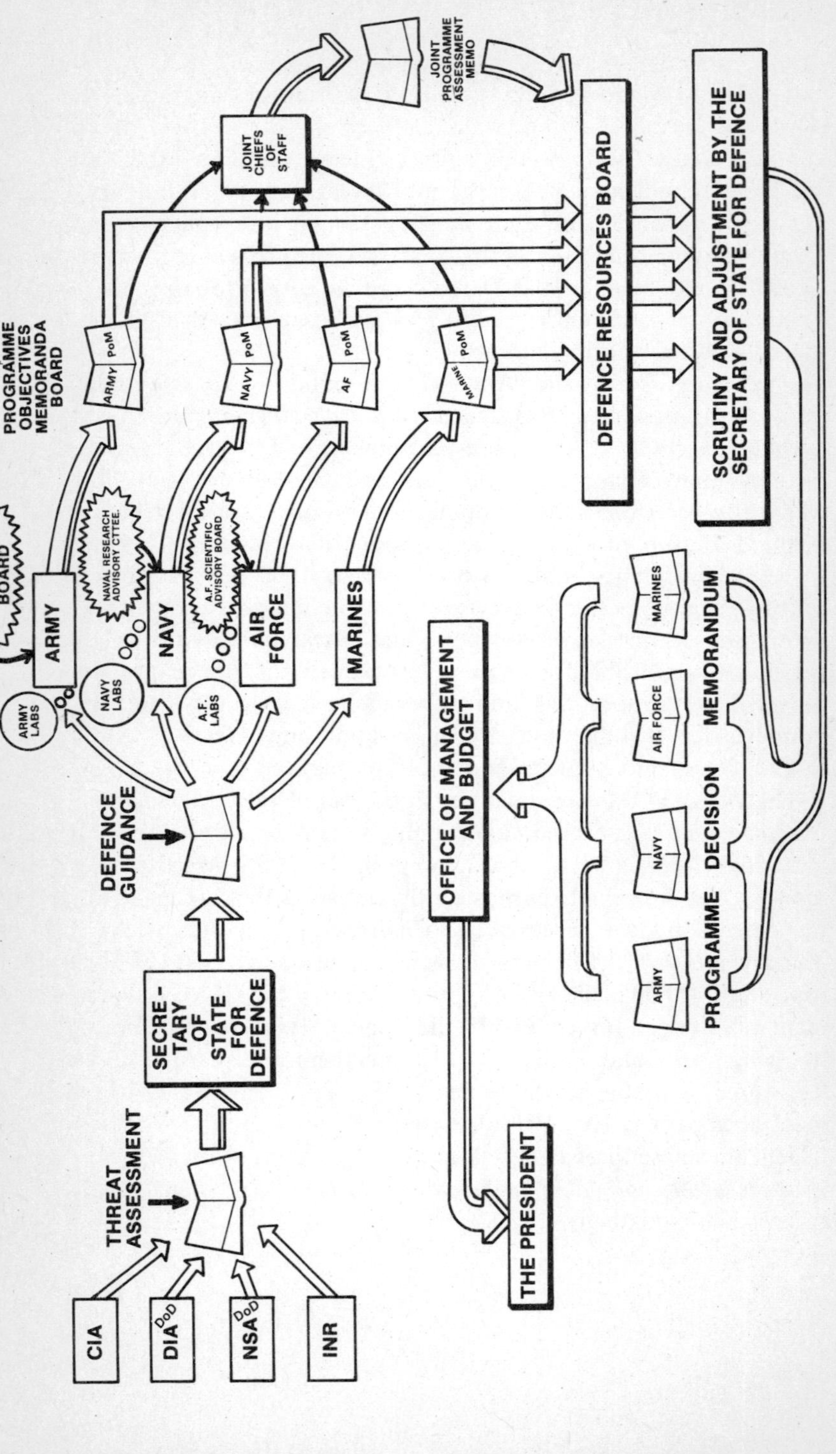

FIGURE 2.5 *Annual planning cycle for delivery systems*

The two main reasons for massive cost overruns are non-competitive contracting and inaccurate costing. A Cost Analysis Improvement Group was set up to meet the latter problem, but in 1983 was still not dealing with all major systems.[18]

Once adjusted and approved by the SECDEF, Programme Objectives Memoranda become *Programme Decision Memoranda*, and move on to the Office of Management and Budget for scrutiny. It is noteworthy that while all these documents are not necessarily classified, they are closely held by DoD and under consideration of 'Executive Privilege'. Congress and congressional staff are denied access to them. This means that Congress must make its funding decisions without knowledge of full programme considerations or options as they have been shaped in DoD.

Even within the DoD, while quite obviously DSARC (weapons-evaluation) decisions have important consequences for defence spending, there are few opportunities within the Pentagon's own decision-making process for the two sets of activities – weapons evaluation and weapons budgeting – to be related. Rivalry between the services, and the inability of the Joint Chiefs to control it, means that the planning process has become 'less an objective analysis of military challenges and possible solutions and more a search for interservice accommodation and after-the-fact rationalisations for favoured weapons programmes'.[19]

This means that there is a shortage of reliable advice for the civilian decision-makers who set military goals. Clear strategy is rare, and hard choices are avoided by embracing a strategic concept so broad that it can justify almost any new weapons system required by the services.

There are also problems at the top – the Secretary and his senior staff are unlikely to be experienced in weapons policy making[20] and their time in office is short (an average of 2.4 years for SECDEFS) seen against the fifteen to twenty years development span for a weapon. 'Those who fathered failures rarely remained in place long enough to take responsibility.'[21]

We now turn to other branches of the executive which may be expected to exercise civilian control over defence decision-making.

At this point it is necessary to examine the different roles of the Joint Chiefs of Staff, and the individual military services, which are in practice an independent source of planning and decision-making, with a major role in the procurement of nuclear systems.

The Joint Chiefs of Staff

Established on the model of the Prussian General Staff, the Joint Chiefs of Staff are responsible for overall military planning. They are a principal source of military advice to the President, and provide strategic direction and operational control of US military forces, including the Strategic Air Command and all US nuclear forces. They prepare both short and long-term military plans, which are intended to guide the defence budget, defence R & D and the contingency plans of operational commanders.

The Joint Chiefs of Staff are limited by law to 400 officers, drawn equally from the Army, Navy and Air Force. There are directorates for Manpower and Personnel, Operations, Logistics, Plans and Policy and Command, Control and Communications Systems. While responsible for speaking with a single military voice, the Joint Chiefs, rather than overcoming service politics, are too often co-opted by it.

> 'Virtually every vivisection of the Joint Chiefs of Staff since 1947 has identified 'dual hat' membership as the root cause of most structural difficulties, but accompanying calls for corrective action have fallen on deaf ears. Severe conflicts of interest still cause co-operative efforts to evaporate under slight pressure and eviscerate options before they can even influence plans'.[22]

They have no direct responsibilities for procurement of equipment, which falls to individual services. Their role in the development of nuclear weapons is consequently less important than that of individual services.

> 'Strategic planning occurs' says Kissinger 'if at all, in the Joint Commands, where the relevant services are brought together for specific missions. But the heads of the Joint Commands neither serve on the Joint Chiefs nor control their constituent elements in peacetime. By contrast, the inevitable and natural concern of the Service Chiefs – with their competitive and often mutually explosive mandates – is the future of the services, which depends on their share of the total budget. The incentive is more to enhance the weapons they have under their exclusive control than to plan overall defence policy.'[23]

The Air Force

With responsibility for Strategic Bombers and the land-based missile force, the Air Force is the leading service in nuclear weapon decision-making. The chart following shows its higher command structure. The US Air Force is organised on a functional basis in the United States, and on an area basis overseas. Of the Major Commands, those with a particular responsibility for nuclear weapons are the Air Force Systems Command and the Strategic Air Command.

The Systems Command is responsible for advancing aerospace technology, adapting it into operational aerospace systems and acquiring new missiles and aerospace weapons. Its budget is about 30% of the total Air Force Budget[24] and it is responsible for the design, construction and purchase of aircraft, missiles and non-nuclear armaments. The Space and Missile Systems Organisation based at Los Angeles is responsible for both developing and testing missiles, which are flight tested from the Space and Missile Test Centre at Vandenburg, California. Programme offices are set up on an *ad hoc* basis under the Systems Command to manage specific missile development programmes. Its Contract Management Division lets the contracts to major contractors, and the Missiles Division works closely with technical experts in the contractor companies.

The Strategic Air Command is responsible for the equipment and operation of the strategic bomber and missile forces, and also for space surveillance and early warning systems. In its mountain bunker at Omaha, Nebraska, and duplicated in continuously airborne command posts, are the control centres which would actually target and launch a nuclear attack. The Strategic Air Command is the main customer in the Air Force for nuclear weapons, and its views have a strong influence on decisions taken at the development and deployment stages. It is a persistent advocate for new and more powerful weapons – for example, it strongly promoted the B-1 nuclear bomber. Even after Carter cancelled the B-1, work continued on R & D programmes on four pre-production aircraft delivered to Edwards Air Force Base, and the Strategic Air Command lobbied hard and successfully for the reversal of Carter's decision when the Reagan administration came to power.

The Navy

The Navy is also an important centre of power and influence over

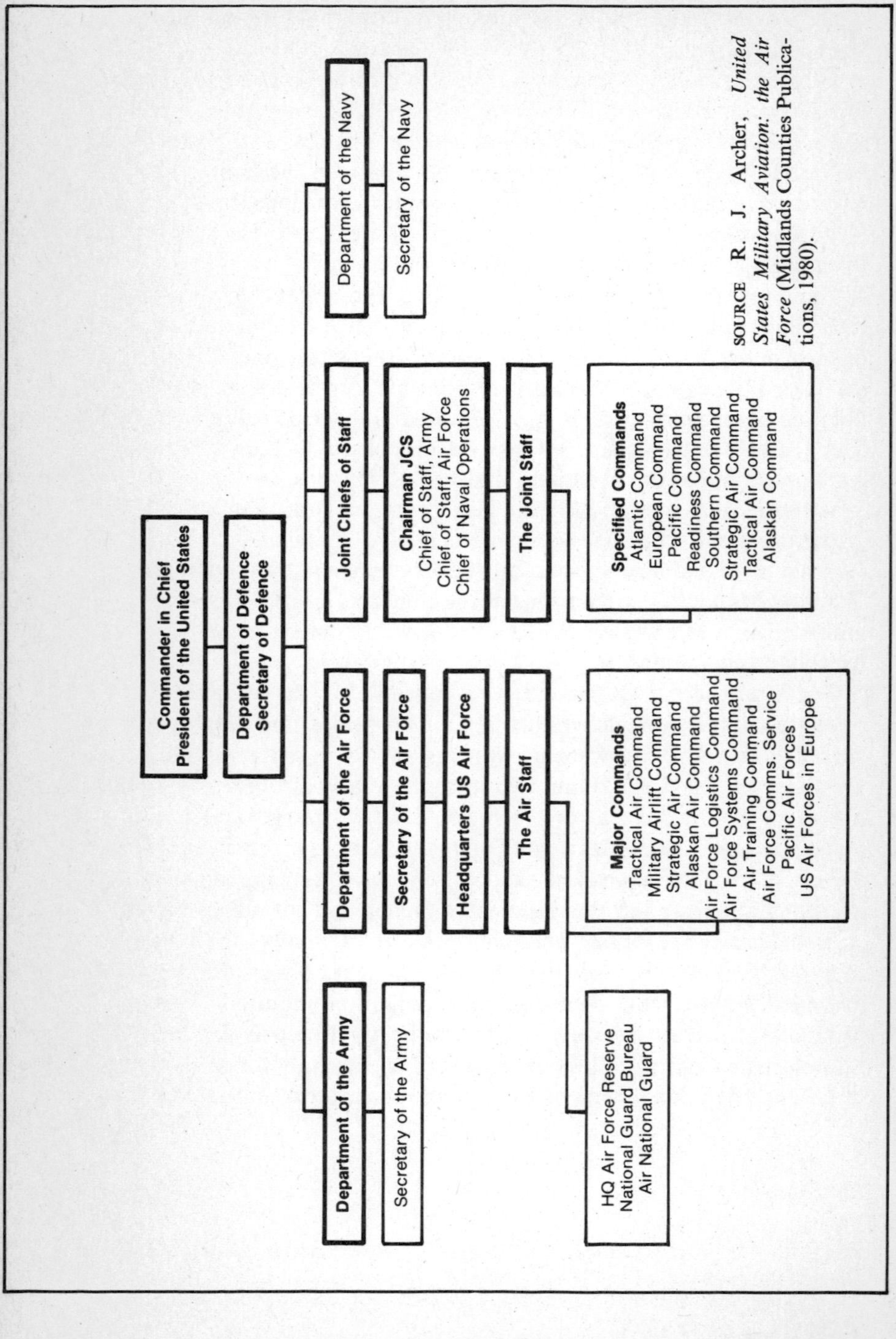

SOURCE R. J. Archer, *United States Military Aviation: the Air Force* (Midlands Counties Publications, 1980).

nuclear weapons development, holding responsibility for the submarine-launched missiles. Like the Air Force, it has specialised divisions responsible for development and procurement of new missiles which act as lobbies for their deployment. As in the Air Force, particular programmes are managed by project teams set up for the purpose, beneath a line of authority which, until recently ran from the Secretary of the Navy at the top, to the Chief of Naval Operations, to the Office of Strategic Offensive and Defensive Systems and the Chief of Naval Material, to the Special Projects Office, which actually administers research and development contracts. This structure is currently being reformed. (US military personnel call organisational charts 'wiring diagrams', and the image accurately suggests the changing connections, bypassing and short-circuiting of departments that occurs as decisions on specific programmes are processed by the military machine.)

The Army

The Army has a smaller role in strategic nuclear weapons than the other two services. Nevertheless, tactical and intermediate-range land-based nuclear weapons are fully integrated into its forces. The top line of authority is the Secretary of the Army, the Chief of Staff and the Deputy Chief of Staff for Research and Development. Within the Army staff, there is a division responsible for Research, Development and Material Acquisition, which conducts R & D, budgeting and procurement of Army weapons, from conception through to acquisition.

An illustration of the independence of the services in nuclear weapons decision-making is provided by the Army's management of the Pershing II programme. The concept for Pershing II, which has an accurate, terminally-guided re-entry vehicle, was put to the Army as an unsolicited proposal by Martin Marietta, the contractor for the Pershing I-A. The programme was approved by the Army and funded by Congress under a research budget line entitled 'Terminal Homing Systems'. In 1978 the Pershing II range was extended to enable it to destroy hardened targets deep within the Western Soviet Union. The deployment programme was speeded up by a year, to meet NATO's 1979 decision to deploy modernised missiles in Europe; as a result, the Army signed a contract for the production of twenty one missiles before development tests were complete and before any missiles had been test-flown. In July 1982, Pershing II failed its test-flight, and in

November the House Defence Appropriations Sub-committee 'deleted' Pershing II procurement funds. Despite this, the Army did not cancel the contract, using other missile programmes' funds to keep the Pershing programme going. By 1983, so much had already been spent that Congress was faced with a choice between cancelling Pershing II, which would cost $300 million, and continuing its funding, which would cost $478 million and provide 91 missiles. Given this choice, and the Reagan administration's desire to deploy the system to put pressure on the Soviet Union at Geneva, Congress allowed Pershing to continue, despite the failure by then of a third of its test flights.[25]

THE ADMINISTRATION

While the Pentagon budget preparation process reflects military requirements for national defence, the President as head of the executive branch of the US government has the ultimate responsibility of weighing up military spending with competing domestic and foreign priorities.

The Office of Management and Budget (OMB)

The President's primary vehicle for making budget decisions is the OMB. Its 600 staff are quartered in the old executive office building adjacent to the White House. Its Director, who is a cabinet member, and Deputy Director are presidential nominees confirmed by the Senate – in fact a significant number of its top posts are political appointments, subject to change with each administration.

While OMB is often seen as a fiscal adversary, even a tyrant, by other departments, its relationship with the DoD is unique. Because of the length and complexity of Pentagon budget-making, OMB has long held that its early involvement in the programming and budget process was essential. It maintains an office and staff at the Pentagon to facilitate co-ordination, and one of its associate directors is a voting member of the Defence Resources Board, thus ensuring some presidential involvement in every step of the DoD's budget process.[26] However, according to March 1983 press reports, twenty-seven examiners review the entire defence budget – that means that each analyst must review over $10 billion in budget authority, twice as much as the average for examiners working on other departmental budgets. There may now be even fewer analysts, representing the fall from grace of the Reagan administration OMB Director after his early battles with Secretary of Defence Weinberger.[27] Indeed, under Reagan, the OMB has played an extremely limited role in defence budget restraint.

The OMB budget schedule runs to tight deadlines throughout the year: one fiscal year's budget is being prepared while the previous one is working its way through Congress. Budget policy development takes place from March to May of each year, and the OMB sets overall guidelines for spending departments; from July to September the departments develop and submit their proposed budgets; from September to December the OMB holds hearings with the departments, analyses their budget, reviews its economic projections and submits the whole to the President. The President decides on totals for each department, but before the OMB prepares the final documents for Congress, the President will hear appeals. The JCS, for instance, is allowed presidential appeal at this stage.

Here as with all budget decisions a yes or no means a programme's life or death, and the battle to keep weapons systems alive goes on through the entire cycle of budget-building. The Pentagon will struggle to clarify its own priorities in order to be in a position to bargain away some programmes to keep others.

Agencies will use friends, allies, and associates from the public sector to bring political pressure to bear to call attention to their concerns.

For example, a potential contractor for a weapons system would contact a friend or associate in OMB to convey the importance of that programme to the economic recovery of the contractor's community. Informal networks are valuable here as are political associates from campaign or fund-raising work. Key Congressmen and women and even Congressional staff are players in this stage of budget-building as they give indications of support or trouble in Congress for certain provisions of the budget.

OMB control over economic projections can also be used to save a programme. In 1983 the B1B bomber was in serious danger of losing the Armed Services Committee appropriations vote on multi-year procurement because of the programme's skyrocketing cost projections. Pressure on OMB from Senate Republican committee members to 'find the money' was great. Rather than sacrifice this programme to the budget knife, OMB issued new figures on the overall economic picture four hours before the scheduled Senate vote. The OMB change allowed projected income figures to match, almost to the dollar, the figure needed to secure the B1B vote.

As the tensions and stakes increase in each year's budget cycle, personal attacks on OMB leadership and staff are not uncommon. As one high official put it, 'Nobody aims for your knees, they aim right here' and he touched the middle of his forehead.

The President and His Staff

Some observers say that the President of the United States is the ultimate decision-maker on nuclear weapons because he sets the tone for policy, articulates that policy to the American people and Congress, and is ultimately responsible for implementing nuclear strategy.

Others say that, in fact, the President only decides what has not or cannot be decided at a lower level. Like other decision-makers, the President will rarely see options unless he insists on them; by the time he is presented with a weapons issue for approval, almost all of the decisions are made.

There is some validity in each of these views. Both the power and the limitations of the presidency are real. Because of his high office and constitutional authority, the President could intervene in nuclear weapons decision-making at almost any point along the line. He is most likely to do so in three contexts: budget consideration, authorisation of the Annual Stockpile Memorandum or arms control negotiating positions. He also has extensive influence through his powers of appointment to senior posts throughout the executive departments, powers more extensive than that of the French president, and certainly much wider than those of the British prime minister.

In his first administration, President Reagan instituted other review processes which he found useful: the creation of special presidential committees to examine nuclear questions, such as the Scowcroft commission on MX deployment and the Fletcher commission for technical study of future nuclear space strategies. These are in addition to his statutory advisory boards on foreign intelligence, arms control and science, the last of which will be examined in some detail later.

The President has the authority to reject a weapon system, but unless he does it at a very early development stage, the potential political cost of such an action is enormous. It means the administration going against the judgment, commitment, prestige and investment of countless experts, military leaders, congress people and contractors. Expending political credits on a weapon decision will mean fewer reserve credits for future battles, as President Carter learned after his B-1 bomber cancellation. Usually a skilled politician, the President will be loath to use those resources unless it is an absolute necessity. He must always consider whether he will be able to keep Congress and the people with him in making such a decision. Moreover, unless he has a comprehensive, alternative policy formulation for national security a single system disapproval or variation would have no logic. If it stands alone, as the B-1 decision did, the dynamic of bureaucratic resurrection takes over and the system will return after that President departs.

The organisation of the executive office differs in each administration and is designed to suit the President's work and decision-making styles. The President must have confidence in the staffs' individual judgments, trust their character and know that they understand his position well enough to speak in his stead. This generally means that the inner circle is made up of those who have survived the political fires along the road to the White House. They usually lack foreign and military policy experience but have reasonably good domestic political sense and experience as well as overall organisational ability.

Running the presidency and the government however are different from running a campaign. The staff must reflect the electoral mandate, so new people must be introduced into the trusted circle. Ideally this inner staff works together smoothly. In reality, personal and ideological differences surface among White House staff and throughout the administration. The inevitable result is bitter bureaucratic infighting and bickering. The President has often had to intervene, for example in Kissinger's fights with Schlesinger, Brezinski with Vance. Other fights within the administration, over MX for example, had to be resolved by meetings at Camp David. Everyone seeks the power of having the President's ear. Lives and friendships shatter in the process and only the most tenacious of the infighters survive. It is this survivability which often determines who makes up presidential staff and that in turn shapes presidential decisions and policy. The President can become a captive of his own inner circle.

The White House Science Council

Presidents since Eisenhower have had a science advisory council of one sort or another. In most cases presidents have had a close personal relationship with and respect for their science adviser, which has in turn facilitated the science community's direct personal access to the President.

In 1974, during the difficult days of the Nixon administration, the President abolished the office after his science adviser resigned in protest at administration policies. Ford never recreated the office but Carter replaced it with the Office of Science and Technology Policy, which remains in place and works in conjunction with the present White House Science Council which was reinstituted by Reagan in February 1982. Its purpose is to advise Dr George Keyworth, Science Adviser to the President, on science and technology issues of national concern. Council members, appointed for a period of one year, serve at the discretion of the President and meet up to six times each year at regular intervals; subgroups of the council may be formed to conduct studies on specific issues assigned by Keyworth, who was previously Director of underground nuclear testing at Los Alamos. The President's passion for the Strategic Defence Initiative is widely thought to have been nurtured and fuelled by Keyworth[28] and his mentor Edward Teller, known as the 'father of the H-bomb'.

President Carter's science adviser, Frank Press, blocked the progress of MX for a year in 1976–7 by questioning the 'tunnel concept' for its basing, and questioning the reality of the Soviet threat to US silos. Later the Air Force and service chiefs used intelligence reports to attack Press' judgment over the Soviet threat.

The National Security Council (NSC)

The White House has the NSC as its own mechanism for decisions on national security. Despite the fact that it rarely meets in formal session, its members are the most critical leaders in foreign and military policy: the President, Vice-President and the Secretaries of State and Defence; the Director of the CIA and Chair of the Joint Chiefs of Staff are statutory advisers. Other participants, such as the Ambassador to the UN, the White House Chief of Staff, and the National Security Adviser, have to be invited. The NSC has top security clearances all round and information and staffing requests are responded to as if it were the President personally.

The NSC however undergoes radical rises and falls in stature as successive Presidents reshape its purpose and structure to suit their temperaments. Its precise workings also depend on who fills the post of National Security Adviser (NSA) and upon the relationship between that appointee and the President. As with other senior White House staff, the NSA works with no statutory description of the position. His appointment by the President is not subject to Senate confirmation.

> In effect, he serves the National Security Council as Executive Director overseeing its dozens of staff analysts and managing the disparate flow of diplomatic and military data coming into the White House from various government agencies. As an assistant to the President, the National Security Adviser has to produce coherent syntheses of this flood of information and opinion, and sometimes arbitrate interagency disputes so that the Chief Executive can make informed, independent policy decisions. And, of course, the adviser must simply advise, telling the President what he thinks ought to be done.[29]

The post has become one of the most powerful in Washington because of presidential access and because of its brokering role: top foreign policy and military staff are available to fill it.

While presidential participation occurs throughout the executive office phase of the defence budget process, with the National Security Council involved to the extent each President sees fit, this involvement peaks in the months before submission of the proposed federal budget to Congress. For FY 1986 the defence budget requested by the President was $277.5 billion.

THE STATE DEPARTMENT

The Department of State is housed in a huge grey building which spreads over four city blocks of Washington's Foggy Bottom area. The plain structure is singularly without grace and conveys the blandness sometimes attributed to the foreign policy professionals who work there. It has been called a 'granite rabbit warren', with each of its eight floors divided into colour-coded wings and recesses which wrap around an inner courtyard. The top floor diplomatic rooms are elegantly appointed with some of the country's finest antiques and artwork. From these graceful environs one looks out over the Potomac river, the sparkling Lincoln Memorial and the sunken, black marble of the Vietnam Memorial.

While the State Department is the official base of foreign policy formulation and implementation, its influence on nuclear weapons decision-making is minor relative to that of the DoD.

The Secretary of State has prime responsibility for negotiating all arms control or disarmament treaties with other governments. Technically, the negotiating task is administered through the Arms Control and Disarmament Agency, which provides and co-ordinates the necessary back-up. The negotiating team may have staff from several agencies, including the CIA, NSA and the Department of Defence. The President appoints the official negotiators who are then confirmed by the Senate.

The Arms Control and Disarmament Agency (ACDA) was created by congressional initiative in 1961 as an independent agency reporting to the President.

Pointing out the intent of the legislation, Senator Claiborne Pell recalls: 'What we intended, 22 years ago, was for ACDA to play the role of an advocate for arms control as a way to enhance our national security. We wanted, in other words, an agency that would more often than not counter-balance what was coming out of the Pentagon.'

Today, ACDA is the glue that keeps arms control functioning on a day to day basis for the government. It fields the negotation delegations, oversees the inter-agency preparation for negotiations, provides delegation background and staffing before and during negotiations and supports the negotiation teams from Washington. Chief negotiators report to the Secretary of State through the ACDA Director.

It can be seen from a quick scan of the official agency outline that the role of ACDA is primarily to prepare and advise; the influence of the agency and its Director are dependent upon the attitude of the President toward arm control issues and the President's relationship with the Secretary of State and the ACDA Director. Similarly, a great deal depends upon personal compatibility between the Secretary of State and the ACDA Director. The ease of working co-operation between Carter appointees Vance and Warnke, for example, enhanced the agency's image and influence within the national security community.

The calibre of agency staff is crucial because much of the policy work around weapons, particularly nuclear weapons, is done in highly flexible formulations known as Inter-Agency or Inter-Departmental Groups. These Groups of officials operate at sub-cabinet level, to consider and co-ordinate policy studies and prepare and review negotiation details. They are chaired generally at Assistant Secretary level, usually from State or Defence. The Group for Europe, which handled long range theatre nuclear force talks and mutual balanced force reductions issues was co-chaired by the Assistant Secretary of Defence for International Security Affairs in the Pentagon, and by the Director of the Bureau of Politico-Military Affairs at the Department of State. The office of the Secretary of Defence may also be represented, the Joint Chiefs-of-Staff, the Military Liaison Committee, the CIA and others, so unless ACDA personnel are of top quality, their perspective may well be swamped.

If a policy paper passes muster at this level it goes to a senior Inter-Departmental Group, of which the Reagan administration has three, for defence, foreign policy and intelligence. A position paper which has been accepted there goes to the National Security Council before the President sees it. The final determination of presentation to the President will be decided between key White House staff and the Group.

ACDA's budget has been cut by thirty per cent since the beginning of the Reagan administration. Most key political appointees are former members of the conservative Committee on the Present Danger formed to oppose the SALT II agreements. Its Research Council has been abolished. The Operations Analysis Division which housed the computer capability, the institutional memory of arms control, has been abolished and the Agency now uses the computer services of the Railway Association. Its centralised records have been delegated to individual bureaus and in the process much has been lost. The ACDA Library has been shipped to George Washington University to be stored in its special books collection, not to be recovered (by agreement) for a minimum of five years.

Instead the President has chosen to set up commissions and study groups all too often to do what the Agency is meant to do. Part of this is an effort to free policy from the mutilating, internal personnel battles between moderate and combative conservative Republicans.

The State Department's *Bureau of Intelligence and Research* has already been described as part of the intelligence community.

The *Office of Congressional Relations* seeks the views of Congress on major foreign policy issues, represents the department in explaining US foreign policy initiatives, and arranges formal and informal meetings, briefings and appearances by department personnel before congressional committees. It co-ordinates with ACDA congressional relations on nuclear weapons issues.

The office is especially active during hearings on nuclear arms control treaties presentations and during discussions and debates regarding overall nuclear weapons policy and strategy. On the Senate side, arms control issues are handled by the full Foreign Relations committee; the House Foreign Affairs committee has an Arms Control sub-committee.

It should be emphasised here that the formulation and implementation of foreign policy is the responsibility of the Executive Branch, and while Congress can advise and criticise, final decisions on overall policy rest with the President.

The *Bureau of Politico-Military Affairs* develops and co-ordinates guidance for US security policy, including military assistance programmes, arms control, nuclear non-proliferation and conventional arms transfer policy. The Bureau liaises with DoD, ACDA and NSA on these issues. Its task is to integrate complex diversities of opinion from within the State Department, other government agencies and public groups for policy recommendations; it is not concerned with specific nuclear weapons.

The *Policy Planning Staff* is the department's centre for policy development relating to nuclear weapons and strategies. Divided into specialist sectors, the Council includes foreign policy and weapons experts from universities, think tanks and State Department professionals. Its work attempts to balance weapons realities with an apparently integrated strategic policy and, more difficult still, with stated foreign policy goals and objectives.

Personnel

As in other American government departments the Secretary of State appoints his own senior department team, but in practice those recommendations must be cleared through the White House personnel office. Political activists who wish to affect the direction of major departments use their influence in this office to move against or for submitted names. In the Reagan administration, for example, the Committee on the Present Danger has been influential in stopping a number of key appointments early on.

Key appointments also require advice and consent from the US Senate, through the Foreign Relations committee. While confirmation hearings are generally perfunctory, it is possible for the Senate committee to use them as a public focus on issues relevant to the appointee's administrative area.

The State Department exemplifies more than most departments, the balance between permanent professionals and political appointees. Historically, the professional foreign service has been drawn from the white male elite trained at the finest institutions of higher education, personifying the 'old boys' network more than most sectors of American life. While there are surface concessions to newcomers, as well as to women and minorities, the older, inner reaches of this circle, often called 'The Club', are not easily permeated. Professional foreign service organisations keep the group together in retirement and between posts.

THE US CONGRESS

Compared to the elected representatives of the other nuclear nations, America's Congressmen and women undoubtedly are far better informed about the nuclear weapons being developed and produced with the monies which they approve.

The US Congress is simple in outline – a 435 member House of Representatives and a 100 member Senate – and enormously complex in its inner workings. The senior members – grand masters of rules, procedures and intra-party politics can wield substantial decision-making influence and power. Most Congress-people can only be generalists given the range of issues with which they must deal, and depend on the experience of staff and experts. Most of the real congressional decision-making, therefore, is done in committee and sub-committee. These committees have much wider powers than their equivalents in Westminster: they can for example, draft legislation or vote on budget amendments, and those votes can pass virtually unchallenged through the full chamber.[30] On defence issues, each chamber has a major committee to deal with authorisation (setting the overall legal limits on spending) known as the House or Senate Armed Services committee, and another 'appropriations' committee to deal with the preparation of the actual spending bill. Each of these has specialist sub-committees on warheads or nuclear delivery systems – see the explanatory table which follows.

Until the mid-1980s the key chairs of these committees, and most of their membership, Democrat and Republican alike was generally considered to be conservative and hostile to arms control. But the election of Les Aspin as chair of the House Armed Services committee in January 1985 completed a 'transformation of four key congressional committees from fortresses of Pentagon protectors to decision-making bodies that might at least occasionally listen to dissenting voices'.[31]

The timetable of the federal budget process, in theory at least, is as follows. In January of each year the President's budget proposal is presented to Congress. This 700-page book is accompanied by a volume of special analyses, and a message from the President, as well as a host of press releases and documents from the spending departments. The administration stages press-packed briefings, and the House and Senate Budget committees begin their hearings, with administration superstars coming to the Hill to present their case and defend their budget. By 15 April the Budget committees draft the First Budget Resolution, and during the next month the full House and Senate debate this resolution. From 15 May until September the Authorisation and Appropriations committees make spending decisions on specific programmes, culminating in the Second Budget Resolution in late September. The new fiscal year begins on 1 October. If Congress has not completed its process before the end of a fiscal year, a Continuing Resolution may be passed by each House.

Congressional staffing and activity on defence issues have expanded dramatically; over two decades the number of hearings of the two Armed Services sub-committees have quadrupled and their professional staff has increased five-fold. Committee members are inundated with data by Pentagon officials, for line-by-line examinations of weapons programmes. The proceedings of congressional committees are televised, and the public can observe the detail of the procedures followed. There are now at least four major congressional institutions to help Congress in its budget work, the General Accounting Office, the Congressional Research Service, the Office of Technology Assessment and the Congressional Budget Office.

Given that so much information is available that Congress has formal control over the purse-strings, and that a growing proportion of US public opinion favours weapons restraint, why is it that, over time,

TABLE 2.1 *Congressional authorisation and appropriation committees*

	House of Representatives	*Senate*
Authorisation: Warheads	Armed Services Committee (Procurement and Military Nuclear Systems Subcommittee)	Armed Services Committee (Strategic and Theatre Nuclear Forces Subcommittee)
Authorisation: Delivery Systems	Armed Services Committee (Procurement and Military Nuclear Systems Subcommittee)	Armed Services Committee (Strategic and Theatre Nuclear Forces Subcommittee)
Appropriation: Warheads	Appropriations Committee (Energy and Water Development Subcommittee)	Appropriations Committee (Energy and Water Development Subcommittee)
Appropriation: Delivery Systems	Appropriations Committee (Defence Subcommittee Military Construction)	Appropriations Committee (Defence Subcommittee)

Congress barely alters what comes from the administration and the Pentagon in nuclear weapons requirements?

This question has received little attention, but at least three major causes are indicated. Firstly, because each weapons appropriation decision is so complex, congressional committees become absorbed in technological detail at the expense of being able to assess the overall defence picture. They have no mechanism for making sure that specific decisions – the W-82 enhanced radiation warhead, the MX missile or the Strategic Defence Initiative – fit into a coherent military strategy.[32] Many of the large, simple questions that the public asks, like what is the real threat, what alternative methods are there of meeting it, are not asked or answered in Congress.

Secondly, committees tend not to pay significant attention to weapons systems when they are in the research and development stages – 'in a budget that is measured in hundreds of billions of dollars, these embryonic weapons systems are poor candidates for close congressional scrutiny'.[33] Yet it is precisely in the early stages that new systems accumulate the key supporters who can see them through full-scale development and mass production, which brings us to the third point. Whereas the overall defence strategy does not attract powerful constituents and critics, and thus major public debate, individual weapons systems do. Each successful one is championed by distinct and powerful constituencies: by that section of the defence bureaucracy responsible for it, by the service whose superweapon it is to be, and by the defence industries whose contracts are at stake. While representatives have to stand for election every few years, each of these communities has tenacity and permanence on its side.

The W-82 155 mm Enhanced Radiation Warhead (ERW)

For sixteen years prior to 1985, the US Army had been trying to build a nuclear artillery shell. In 1978 President Carter decided not to deploy new enhanced radiation warheads for the Lance missile. In 1983, when the subject came up before the House Armed Services committee, the members directed that further research and development on the warhead be terminated. Meanwhile the Energy Department continued to spend $21 million on its development in 1984, keeping the programme afloat, but omitted funds for another ERW weapon, the W-82 artillery shell, in its budget for fiscal 1985.

The House Military Nuclear Systems sub-committee chair, Samuel Stratton, an avid supporter of battle-field nuclear weapons, was very upset: 'To clarify matters the sub-committee's counsel revealed that there appeared to be some flexibility in the Energy Department's R & D budget which it was not taking advantage of'. A week later General Bernard Rogers, Supreme Allied Commander Europe, made a forceful appearance before the full House Armed Services committee, outlining the role of the new shell in the strategy of flexible response. Rogers argued that the W-82 warhead be produced in such a way that it could become enhanced-radiation capable by the insertion of a tritium module. 'Keep the modules over here, and when the time comes that it's decided we'll make enhanced radiation, and the allies have no gas pains over that, then we can send the modules over.'

Richard Wagner, Assistant to SECDEF for Atomic Energy, testified to Stratton's committee on deployment of the neutron warhead in February 1984, bringing along a Marine Corps major to buttress his case. On 21 June 1984, the Senate was staggering through an all night session to complete the massive $230 billion defence bill; in an undebated amendment at 3.00 a.m. it re-authorised $50 000 000 production funds for the new nuclear artillery shell.[34]

DEFENCE CONTRACTING

The story of the W-82 warhead illustrates some of the influences over congressional decision-making; we now turn to another – that of the major corporations.

A larger and larger share of the Pentagon budget is being spent on hardware – the development and production of weapons of all types. In fiscal 1980 it was 36%; by fiscal 1983 the proportion had risen to 45%; if current budget plans are approved, the House Budget committee projects that it will reach 52% by fiscal 1988. This rising share of the Pentagon budget for modernisation is at the expense of what is known as 'readiness' (maintenance, training, personnel) and means that the DoD may be swapping today's preparedness for tomorrow's hardware. It also means increased contracts for already gigantic defence corporations.

In nuclear terms, orders for weapons and delivery systems are concentrated among a few companies at the very top of the list. In fiscal 1982, for instance, 82 per cent of the $6.6 billion prime contract awards for major nuclear delivery vehicles were won by the top eight contractors. (Up to half of the contract may subsequently be sub-contracted to some 10 000 other companies involved.) Table 2.2 high-lights the value of nuclear weapons orders to these corporations and the high dependence some of them, notably Rockwell and Martin Marietta, have on nuclear business.

The US Census Bureau released figures in March 1985, based on the annual reports of ten of the largest weapons makers (including those in the chart except General Electric), showing that, on average, they realised a 25% return on equity in 1984. Lockheed was as high as 42%. In contrast, the average return for manufacturing corporations of all sorts was only 12.8%. (The measure of profitability which was used is one widely accepted on Wall Street, and accepted as fair by the companies.) The ten companies also reported that their backlog of government orders, an indicator of future profits, totalled more than

$80 billion. Their backlog of commercial business is considerably smaller.

As profits of the industry have increased, so has the questioning of that profitability. In March 1985 the Air Force requested that two contractors return $208 million in excess profits; one of these was General Electric, temporarily suspended from obtaining new contracts following a federal Grand Jury indictment charging the company with falsifying $800 000 in claims and lying to the government about work

TABLE 2.2 *Dependence of top nuclear contractors on nuclear delivery weapons contracts*

Contractor/project	*Total nuclear delivery awards $million fiscal 1982*	*Nuclear awards as % total sales*
Rockwell International	1 644.5	21.6
B-1 bomber	1 299.9	
MX surface-launched missile	344.6	
Boeing	875.2	9.7
B-52 bomber	488.4	
Air-launched cruise missile	95.9	
Grd/sub-ditto	13.0	
B-1B bomber	277.9	
Martin Marietta	803.6	22.8
MX surface-launched missile	452.3	
Pershing surface-ditto	351.3	
Lockheed		
Trident sub-launched missile	776.5	13.8
General Electric	350.7	1.3
MX surface-launched missile	50.0	
Grd/sub-launched cruise missile	0.3	
B-1B bomber	256.6	
Trident sub-launched missile	43.4	
General Dynamics		
Grd/sub-launched cruise missile	341.8	5.5
McDonnell Douglas		
Grd/sub-launched cruise missile	217.8	2.9
Northrop		
MX surface-launched missile	214.0	8.6

SOURCE: US Department of Defence.[35]

on a warhead for the Minuteman missile. In the same month the Pentagon placed a thirty day freeze on overhead payments to General Dynamics, pending investigation of $244 million in alleged excess charges and unexplained expenses billed to the Pentagon including such items as golfing weekends and kennel fees for a company executive's dog. There have been seven separate government investigations into the performance of General Dynamics, the company which is helping to design Britain's Trident submarines. In 1984 Boeing charged the Pentagon $127 000 for political contributions made to politicians, and $36 000 of 'community relations' expenses including a golf tournament for scouts. Pentagon auditors have raised questions about millions of dollars of such expense claims filed by Lockheed, McDonnell Douglas, United Technologies and Rockwell International. Nine of the ten largest Pentagon contractors, and 45 of the largest 100, are under criminal investigation.

These mounting allegations reflect the unease of a nation whose economy has grown heavily dependent on its arms industry, where the Pentagon is the largest single purchaser of goods and services, where defence industries account for one tenth of all manufacturing; a nation warned by President Eisenhower in 1961:

> In the councils of government we must guard against the acquisition of unwarranted influence, whether sought or unsought, by the military industrial complex. The potential for the disastrous rise of misplaced power exists and will persist [and if not heeded] the nation will be robbed of its greatness by the inexorable demands of a system designed to keep it free and vibrant.

The influence of defence industry on arms production is undoubtedly stronger in the US than in any other nation but its actual extent is subject to debate. Defence contractor influence on Congress is limited by the fact that congressional control over defence spending is limited. By the time a weapons expenditure is large enough to warrant congressional attention, chances are that the system has outlived its competitor weapons within DoD. There are, however, significant instances of contractors swaying a key vote by lobbying of Senators and Representatives in whose 'home districts' sub-contracts have been judiciously placed. At the time of the March 1985 vote to fund the MX missile, for example, a pattern of sub-contracts was outlined covering nearly every state.

Another useful method of unlocking congressional doors is campaign contributions from contractor-sponsored political action committees (PACs). According to Federal Election Commission records, the top 18 defence contractors disbursed through their PAC's almost $2.4 million to congressional candidates in the two years prior to the 1982 congressional elections. Most of these contributions went to members on key funding committees: a Common Cause analysis of contribution data for the 1981–2 election cycle shows that members of the House of Representatives Appropriation Defence sub-committee received an average donation of almost $16 000 compared to an average of just over $2800 for the House as a whole.[36]

In terms of overall influence on weapons decisions, the defence industry's pressure on and co-operation with the Pentagon is much more significant. Multi-million dollar investments in research and development on new weapons concepts, often but not always subsidised by the Department of Defence, means that defence corporations have a powerful interest in selling their concept in its very early stages to the armed forces. Once awarded, these weapons contracts all but guarantee that the competition on that system fades and DoD is a dependent customer. This in turn has spurred the practice known as 'buying in' by which a company bids low and the system looks do-able and affordable. Then the escalation costs and adjustment expenses begin, but there is no way out as the contract is locked in and the weapon has been declared essential to national security. There is a 91% chance of a weapon cost overrun that will average 52% of the weapon's cost. Competition has so little place in the DoD's dealings with its contractors because both sides find its absence convenient. The contractors are freed from market forces and the Pentagon's managers are free to change and gold plate weapons at their whim.

The day-to-day, side by side work between parallel corporate and government weapons programme offices facilitate a close working relationship which nurtures the development of a weapon. Both the corporation and the government offices have an investment in the weapon's success and that common bond, above all others, makes the co-operative relationship an almost natural one. This spirit of co-operation is not surprising, given that many of the individuals involved in defence spending issues have moved back and forth between positions at DoD and jobs with the defence industry. Between 1970 and 1979 over 1600 people transferred from senior posts in the Pentagon and NASA to positions with one of eight major defence contractors. During the same period, 223 executives moved in the reverse direction.[37]

THINK TANKS AND PRESSURE GROUPS

The pressures on nuclear weapons decisions come not only from the military and civilian personnel of the Department of Defence, the Department of Energy laboratories and the major hardware contractors, but also from a large number of defence analysis institutions known as the 'National Security Community'. Starting from a handful of specialists in the early days of the Rand Corporation, this peculiarly American blend of civilian expertise with government money now produces a barrage of studies and recommendations for the use of military power. The 'think-tanks', as they are known, may be policy institutes like the Heritage Foundation, the Hudson Institute, the Brookings Institution, or the American Enterprise Institute, supplied by foundation grants or government subsidies for particular studies; they may be university departments, such as Georgetown University's Centre for International and Strategic Studies, or the Centre for Foreign Policy Development at Brown University; they may be Federal Contract Research Centres, controlled and funded by the federal government, such as the Centre for Naval Analyses, the Institute for Defence Analyses, the Aerospace Corporation, Mitre Corporation and Rand; or they may be private consultancies run on a profit-making basis working under contract to the Pentagon, described in 1985 as 'sprouting like mushrooms all over Washington'.

The importance of these organisations lies not so much in the studies they produce or in the war-games they play, as in the fact that they have developed such a highly esoteric expertise that the community has been compared to a brotherhood, one whose members revolve in and out of high administration positions. We have already noted the importance of the President's National Security Adviser; the same syndrome of influential advisers is duplicated throughout the Pentagon and State Department, with experts from think-tanks being called in on specific questions, and offered senior positions with changes in administration. 'Think-tank reports are considered a preview of what might happen after the next election. The author of a particular report might become an Assistant Secretary in the State or Defence Departments'.[38] As the jargon of strategic debate and the arcane acronyms of weapons systems drive discussion into more and more compartmentalised and specialist fora, these analysts, wargamers and consultants are tending to set the parameters of the defence debate. 'These questions are too complicated for either politicians or the public', said a former national security official in the Carter administration, 'They need help'.[39]

DECISION-MAKING ON THE MIRV–A CASE STUDY[40]

Like most of the major innovations in the nuclear arms race, MIRV – the technology of placing multiple, independently targetable warheads on missiles – was developed first in the USA. By permitting one missile to attack several missiles, it now marks an important step towards 'counterforce' weaponry. It provides a clear illustration of the interplay of technological momentum, bureaucratic manoeuvring, strategic doctrines and perceptions of Soviet threats which typifies the development of a major new weapons system in the US.

The key decisions to develop MIRV were taken in the mid-1960s, in the early period of the Vietnam war. MacNamara was the Secretary of State for Defence and Kennedy was President.

MIRV was conceived in 1962, in the weapons laboratories and aerospace industries of Southern California.

The concept behind MIRV was to place a manoeuvrable 'bus', containing several warheads, on top of a missile. The bus would detach itself from the missile's rocket outside the atmosphere and launch a number of warheads, one by one, on independent ballistic trajectories.

Engineers at four defence contractors discovered the concept independently – Aerospace Corporation, Autonetics and Rocketdyne (a division of North American Aviation), Space Technology Laboratories and the Lockheed Missiles and Space Company. Physicists in the Physics Department of the Rand Corporation also arrived at the idea at the same time.

The defence contractors enjoyed close relationships with the services (especially the Air Force) and the Department of Defence. Staff moved freely between the Air Force Ballistic Missiles Division at Norton Air Base, San Bernardino, and the laboratories of the contractor companies. They were engaged in a constant process of refinement and development of missiles and new weapons; they competed with one another for contracts, and their funding and their ability to attract top staff depended on keeping at the leading edge. Many key developments were produced by the engineers themselves, responding to the challenges thrown up in the process of researching and developing new missiles. In the case of MIRV, there was no 'mission requirement', no specification for the new weapons set by strategic doctrine. The concept arose as an evolutionary development from earlier design work on guidance and propulsion systems and multiple warheads. Only Albert Latter, from RAND, connected the idea of MIRV with ICBM vulnerability, and his view that the US Minuteman force was vulnerable to

Soviet missiles at that time was not widely shared. No elected officials or political accountability were involved in this early stage.

Once it had been invented, MIRV had to be 'sold' to the Department of Defence. The key divisions in the services were the Ballistic Missiles Division in the Air Force and the Special Projects Office in the Navy. Both were responsible for new missile developments, and both depended on new missiles being accepted for their funding and status. The Ballistic Missiles Division had particularly close contacts with the defence contractors, and needed little persuading that MIRV was technically a 'sweet' idea. To communicate to higher levels of the bureaucracy, the scientific advisers from RAND and the Lawrence Livermore Laboratory played an important role, using their positions on scientific advisory boards to various departments in the services and the Department of Defence to present the case for MIRV. Here Albert Latter, Edward Teller and Carl Hausman were influential. They all knew each other personally, and were also well known to Harold Brown, the head of the Livermore Laboratories since 1961, now in the key post of Director of Defence Research and Engineering.

In this stage of its development the key decisions were made by a limited number of people from the Ballistic Missiles Division, the Special Projects Office and the Director of Defence Research and Engineering. The latter was the only politically accountable official to be fully involved. MacNamara learned of MIRV in 1964 and approved the letting of the early contracts; he was, as Secretary of State for Defence, unusually immersed in the details of defence contracting. The Office of Systems Analysis in the Department of Defence, the main civilian department concerned with strategic defence planning, did not consider MIRV until 1965, by which time a strong consensus in favour of MIRV had developed in the concerned Service Departments and the Directorate of Defence Research and Engineering.

It was also necessary to win the support of the rest of the Air Force and of the Navy, especially that of the Strategic Air Command. Curtis LeMay, the champion of strategic bombing and planner of the fire bomb raids on Japan at the end of the war, was still the commander of SAC. At first he was sceptical of the claims made for the accuracy of MIRV and preferred instead to build many more Minutemen. But when the scientists proved their case for accuracy, and it became clear that the Air Force was not going to get the huge force of Minutemen it wanted, LeMay – and the rest of the Air Force – swung behind MIRV. The Navy was more resistant at first, the operational division fearing that the Poseidon missile would swallow resources they wanted to spend on ships. Poseidon was delayed for a year, but once it had been accepted, the Navy did not resist MIRVing the missile.

The key decision in MIRV's history was to begin active development work; this was taken in 1964 by the Director of Defence Research and Engineering, Harold Brown. Although only a commitment to develop an option, this decision gave MIRV considerable mementum. Brown later told the House Appropriations Committee:

> The decision to deploy is not yet made. But there is a strong presumption that it is a likely decision on the basis of our decision to proceed with a $1 billion development. My own view is that you do not proceed with a $1 billion development unless you think there is a high chance of deployment, and that is the situation.

The decision to proceed was made with reference to the strategic doctrines then current. But the connection between ruling strategic doctrine and the procurement of new weapons is not necessarily close, even though the development of a weapon is rationalised in strategic terms. The strategic doctrines used by successive US Defence Secretaries to plan the deployment and targeting of nuclear weapons have undergone a series of shifts and changes of course in the period since 1945. At times the Air Force and the Navy have held their own doctrines, not necessarily compatible with the Defence Department's, and members of the scientific and technical community responsible for developing and promoting new weapons hold a variety of views of their own. In 1964, the Air Force believed in the doctrine of counterforce; the Navy believed in massive retaliation against cities; and MacNamara was in the process of changing official policy from 'counterforce' to 'assured destruction'. The advocates of MIRV were able to present the missile to the Air Force as a counterforce option that would enhance the power of the offensive and to MacNamara as a missile which would offer more assured destruction after a Soviet first strike.

For MacNamara, however, the key factor was that MIRV fitted in with his plans to cut back on manned bombers and to curb the growth of the land-based missile force. By putting multiple warheads on missiles, he could contain pressures for an ever-increasing missile force, for which he could see no political utility, while also keeping the Air Force in rein.

As early as 1958, multiple warheads had been seen as a way of overcoming possible Anti-Ballistic Missile (ABM) defences. At that time the Soviet Union was known to be working on ABM systems.

Kruschev had boasted in 1962 that Soviet anti-missile missiles could 'hit a fly in outer space'. Although there was no firm evidence of Soviet ABM deployment, the fact that MIRV could overcome an ABM system was one of the features which helped to win its easy acceptance. Since new weapons systems take more than ten years to develop and deploy, US military planners have to guess what their Soviet counterparts will deploy in the same period; and the tendency is always to make 'worst case' assumptions. Intelligence reports from 1964 onwards suggested that the Soviet Union was constructing an ABM system, the 'Tallinn Line'. This provided a further rationale for MIRV; and it influenced particularly the planners in the Office of the Secretary of Defence. After 1967 it was appreciated that the 'Tallinn Line' was an air defence system with no signficant ABM capability.

It is likely, however, that MIRV would have proceeded even without an identifiable Soviet ABM. Dr John Foster, the Director of Defence Research and Engineering from 1965 to 1973 said:

> Our current effort to get a MIRV capability on our missiles is not reacting to a Soviet capability so much as it is moving ahead again to make sure that, whatever they do of the possible things that we imagine they might do, we will be prepared . . . we are taking action when we have no evidence (or very little evidence) on the other side of any such actions.

> MacNamara took the decision to deploy MIRVed missiles in 1965 as part of the FY 1967 budget planning. MIRV did not have to be approved by Congress in its own right, since as a delivery system it was presented as a component of the Minuteman III and Poseidon missiles. The funding of these missile programmes were approved annually by the Armed Services and Appropriations Committees of Congress without dissent.

Controversy began only in 1968, by which time it became known that the Soviet Union did not have a fully fledged ABM system, and SALT talks were about to begin. The US decision to deploy the Sentinel ABM system had been taken in 1967. There was widespread opposition to involvement in Vietnam, and interest in arms control was growing.

But within the government, the opposition was muted. In the

preparation of bargaining positions for the SALT talks, the State Department and the Arms Control and Disarmament Agency were agreed that limiting the ABM was the major issue. It was made clear that the Joint Chiefs of Staff would oppose any attempt to put MIRV into the SALT bargaining. The Arms Control and Disarmament Agency therefore took a tactical decision not to resist MIRV, concentrating instead on ABM. Only Morton Halperin, the Deputy Assistant Secretary of State for Defence, protested seriously, when MIRV was about to be tested in 1968. He argued that the tests would remove the option of including MIRV in a missile treaty. Senators Cooper and Hart independently wrote to President Johnson making similar points. But the Joint Chiefs of Staff, the Director of Defence Research and Engineering, the Air Force and the Navy all wanted to exclude MIRV from SALT; better to deploy it first, they argued, and bargain from a position of strength. This view prevailed and MIRV was not included in the talks.

MIRV again became a matter of controversy in 1969, in the early period of the Nixon Administration, the issue centring around verification. A panel was set up under Kissinger, but it did not recommend a ban. In the Senate, the 'doves' who were critical of MIRV could not come close to mobilising a majority, and, like the Arms Control and Disarmament Agency they decided to concentrate on resisting ABM. Criticism expressed in public by advocates of arms control and in the press had no influence on the decision-making process. The opposition, such as it was, had come too little and too late.

The first group of ten MIRVed Minuteman III missiles became operational in June 1970; the first patrol of a submarine armed with MIRVed Poseidon missiles took place in March 1971. The Soviet Union conducted its own first flight-test of a MIRVed missile in 1973.

NOTES

1. For example, Air Force Electronic Security Command, Air Force Intelligence Service, Air Force Technical Applications Centre, Strategic Air Command, Naval Security Group Command, Naval Intelligence Command.

2. Figure 2.1 is drawn from: J. M. Collins, *US Defence Planning – a Critique* (Boulder, Col.: Westview Press, 1982) and W. V. Kennedy, *The Intelligence War* (Salamander: London, 1983) pp. 39–40.
3. Reported in the *International Herald Tribune* (4 Jan. 1985).
4. In February 1985, the Defence Intelligence Agency estimated an increase of 5–8% in Soviet procurement of major weapons systems in 1983. The Central Intelligence Agency, in congressional testimony, referred to 'a stagnation in spending for military procurement after 1976 lasting at least seven years' conceding only the possibility of a modest growth of at most two per cent in 1983.
5. In 1984, the House Permanent Select Committee on Intelligence led a congressional effort to cut off all funding to the CIA-backed rebels fighting the government of Nicaragua.
6. House Appropriations Committee, *Energy and Water Development Appropriations for 1985*, part 6, 98th Congress (13 Mar. 1984) p. 10.
7. T. B. Cochran, W. Arkin and M. Hoenig, *Nuclear Weapons Data Book*, vol. I. (Ballinger: 1984) p. 14.
8. *Philadelphia Enquirer* (28 Oct. 1979) p. 3G, and *Bulletin of the Atomic Scientists* (Apr. 1984) p. 40.
9. 'Sandians Help Define New Defence Initiative', *Sandia Lab News*, vol. 35, no. 22 (28 Oct. 1983).
10. House Armed Services Committee, *DoE Authorisation Legislation for FY 1983*, 97th Congress (26 Apr. 1982) p. 8.
11. D. C. Morrison 'Energy Departments Weapons Conglomerate' in *Bulletin of the Atomic Scientists*, vol. 41, no. 4 (Apr. 1985) p. 32.
12. Draft Report, Centre for Strategic and International Studies at Georgetown University (Jan. 1985).
13. *Organisations and Functions Guidebook*, Directorate of Organisational and Management Planning (OSD: May 1984).
14. M. Rovner, *Defence Dollars and Sense* (Common Cause, 1982) p. 23.
15. Report of Pentagon Inspector General 1984, quoted in R. J. Smith, 'Pentagon Decision Making Comes Under Fire', *Science* (4 Jan. 1985) p. 32.
16. Letter from J. F. Ahearne, Resources for the Future, in *Science* (Feb. 1985).
17. Hearings of Oversight sub-committee of Senate Committee in Governmental Affairs (Jan. 1985).
18. Senate Committee on Governmental Affairs *Management of the Department of Defence* (23 Mar. 1983) p. 17.
19. Rovner, op. cit., p. 38.
20. J. M. Collins, *US Defence Planning: a Critique*, introductory statement for Hearings on Defence Management by the Senate Governmental Affairs Committee (Mar. 1983) p. 159.
21. Collins, op. cit., p. 159.
22. J. M. Collins, *JCS Planning Apparatus* (Congressional Research Service, The Library of Congress, May 1982).
23. 'Arms Debate Must Link Policy and Technology', *International Herald Tribune*, (4 Mar. 1984.)
24. R. J. Archer, *United States Military Aviation: the Air Force* (Midland

Counties Publication, 1980) p. 57.
25. This example of the Pershing II development is based on an article by C. Paine in the *Bulletin of the Atomic Scientists*, (Nov. 1983) pp. 6–9.
26. Rovner, op. cit., p. 25.
27. Rovner, op. cit., p. 47.
28. *Observer* (27 Mar. 1983); *Washington Post* (26 Mar. 1983); *Financial Times* (27 Mar. 1984).
29. *Time Magazine* (Nov. 1983).
30. Floor challenges are however becoming more frequent. From 1968–74 the Armed Services Committees' authorisation bill received an average of more than twenty amendments annually.
31. J. Issacs, 'New Lineup On Defence Committees', *Bulletin of the Atomic Scientists* (Mar. 1985).
32. J. N. Bearg and E. Deagle, 'Congress and the Defence Budget', in Endicott (ed.), *American Defence Policy* (Johns Hopkins University Press, 1977).
33. Rovner, op. cit., p. 63.
34. This account is drawn from C. Paine, 'Senator Nunn's Shell Game', *Bulletin of the Atomic Scientist* (Feb. 1985).
35. If this table also included dual-capable systems, such as F-15 and F/A-18 (McDonell Douglas) and F-16 (General Dynamics), the percentage figures would be higher.
36. M. Rovner, op. cit.
37. G. Adams, *The Politics of Defence Contracting: the Iron Triangle* (Transactiòn Books, 1982) pp. 77–8.
38. E. Schottle, International Affairs Programme Office at Ford Foundation, quoted in *New York Times* (3 Mar. 1985).
39. Quoted in *New York Times*, (3 Mar. 1985).
40. This case study is based largely on an academic study by T. Greenwood, 'Making the MIRV: a Study of Defence Decision-Making' (Mass.: Ballinger, 1975).

3 Britain

This chapter is organised according to the main influences on nuclear weapons decision-making in Britain. Although it starts with the Cabinet and Cabinet Office, the Cabinet is by no means the sole decision-maker on nuclear matters in Britain which many assume it to be. This will become clear as the respective influences of the British civil service structure, the Ministry of Defence itself, the Treasury and the Foreign and Commonwealth Office are discussed. Britain's 'special relationship' with the US on the nuclear issue is clearly a major influence on decisions, and is dealt with next, followed by a brief examination of the importance of Britain's role in NATO. The case-study chosen for this chapter describes decision-making on the Chevaline warhead, an improvement to the Polaris system undertaken in the late 1970s, and which became operational in 1982. This subject was chosen for two main reasons. It illustrates some of the influences on decision-making described in earlier sections, and much less has been written about it than about other British nuclear systems.

It may come as a surprise to some that the role of Parliament is not sufficiently important to appear on the list of major influences on nuclear weapons decision-making in Britain – its role is indeed described under the first section, but its lack of influence signals a theme which recurs in the accounts of the influence of various bodies, and that is the question of accountability in nuclear decision-making in Britain.

THE CABINET AND THE CABINET OFFICE

The Power of the Prime Minister and Cabinet Committees

Decisions on defence and foreign policy are usually assumed to be made by the Prime Minister and Cabinet. The Cabinet, however, by no means always decides on particular questions as a whole Cabinet. Decisions can be taken in sub-committees, which bind the full Cabinet through the concept of 'collective responsibility', although other ministers may be ignorant that such a committee even exists.

These committees are the vital working parts of the engine room of the real machine of government at Westminster, and very little is known about them.[1] In 1984 there were thought to be some 25 known as Standing Committees, and about 110 of the even more secret *Ad Hoc* Committees. This is considered quite trim by post-1945 standards.[2]

Standing Committees are more permanent and slightly less obscure than the *Ad Hoc* Committees. Margaret Thatcher acknowledged the existence of four: the Standing Committees on economic strategy, home affairs, legislation, and overseas and defence.[3] This last is usually known as the Defence and Overseas Policy Committee, and includes the Prime Minister, the Chancellor of the Exchequer, the Home Secretary, the Secretaries of State for Trade and Industry, for Employment and for Defence, and the Foreign Secretary. The Chief of Defence Staff may also attend. There is no rule about how frequently this committee should meet, and its functioning depends largely on the personal style of the Prime Minister. Under James Callaghan's leadership the committee dealt with its business by meeting and discussion; under Margaret Thatcher, the style of communication is largely by written memo, as became clear in the events leading up to the Falklands crisis.

Ad Hoc Committees are set up to discuss a specific subject and disbanded once the matter has been decided. They used to be identified by the prefix GEN and a number (the original decision to make a British bomb was taken by Cabinet sub-committee GEN-163); they are now known by the prefix MISC (for miscellaneous). Tremendous secrecy surrounds the existence of these committees; the rare news of them which emerges is through the memoirs of former prime ministers, or occassionally in the work of journalists. The only Ad Hoc Committees dealing with defence issues currently known to exist are MISC 7 (replacement of the Polaris force with Trident) and MISC 91 (choice of ALARM anti-radar missile), both chaired by the Prime Minister; MISC 32 (deployment of the armed forces outside the NATO area); MISC 42 (military assistance for the armed services of friendly powers), and MISC 51 (commodities needed for strategic purposes), all chaired by a Cabinet Office official.

The Ad Hoc Committee MISC 7 was one of the first Cabinet subcommittees set up in 1979 by the Conservative Government, consisting of Margaret Thatcher, the Home Secretary, the Chancellor, the Defence and Foreign Secretaries. They decided to replace Polaris with

Trident for strategic use well into the 1990s. The internal MoD debate that followed was not about policy but about whether to opt for the smaller Trident C4 or the larger D5 missiles. The argument was settled in favour of D5, increasing the UK strategic weapons capacity from the Polaris fleet with a targeting capability of sixty four to Trident D5 with up to 896 warheads.

MISC 7 was essentially a continuation of James Callaghan's ultra-secret committee set up in January 1978 when the Ministry of Defence and the Foreign Office sought ministerial approval for studies on possible successors to Polaris. Civil Service working parties of officials drawn from these two ministries, the Treasury and the Cabinet Office had prepared the necessary papers, but the issue was deemed much too sensitive even to be put before the Cabinet's full Defence and Overseas Policy Committee. Instead a 'committee of four' was set up to include James Callaghan, Denis Healey (Chancellor), David Owen (Foreign Secretary) and Fred Mulley (Secretary of State for Defence). By the time Parliament was dissolved, the committee of four had reached no firm conclusions but certain things had become clear. The intention was to proceed to a third generation of nuclear weapons despite what had been promised in the 1974 Labour Party manifesto.[4]

Two important issues arise here. One is the role of the civil service in preparing the briefs upon which these decisions are based – the 'working parties of officials' mentioned above – to advise ministers on the political, military and technical implications of the decision. This civil service role is of key importance and will be discussed in detail in the next section. The second is the pervasive nature of Cabinet secrecy in the context of nuclear defence policy. Not only is the information upon which decisions are based kept entirely secret, Cabinet papers being protected from disclosure for thirty years – or even longer, in the case of nuclear decisions, by discretion of the Lord Chancellor.[5] But the decision itself may not be announced to Parliament for many years. Information necessary to enable Parliament and the country as a whole to debate the issue is withheld. Decisions which significantly affect not only defence but foreign policy and the national budget, are taken by four or five ministers, without discussion by the full Cabinet, and without any occasion for debate in Parliament.

The Role of Parliament

Under British parliamentary tradition, Cabinet Ministers are accountable to the House of Commons in certain ways: answering acceptable parliamentary questions, replying to letters from MPs, taking part in debates. But the actual *control* that Parliament could exercise over defence policy would be through two avenues.

Special Debates

The possibility is always open to the opposition to call for a debate on government policy. If the government majority is small, and if the opposition can win over government backbenchers, it can pass a vote of No Confidence. If that happened the government would normally be forced to resign. In practice this has not happened on nuclear issues, for two reasons. Firstly, because insufficient information has been made available to enable a fully-fledged debate on crucial decisions, and secondly because the Labour Party in opposition has until recently been deeply split on the issue. The Conservative Party has had a substantial majority, and in opposition would have been uninterested in calling a debate on nuclear weapons. For thirteen years prior to 1980 there was indeed no parliamentary debate on nuclear weapons at all, a period during which plans were laid for massively increasing Britain's nuclear capability.

The Defence Vote

The Defence Estimates are made public in a White Paper, usually in January each year, and voted upon. The figures in the defence budget are set out under many different headings: Expenditure on Equipment, Expenditure on Personnel, even Expenditure on Research and Development and on Nuclear Strategic Forces. However, nowhere do Members of Parliament see what American legislators see, namely the 'line items' which describe the cost of each weapon separately, in its research, development, testing and manufacture stages. Only in that way would they be able to recognise and to control the development of new nuclear weapons.

The development of the Chevaline warhead for Polaris illustrates this fairly well. In 1967 Aldermaston was asked by the Wilson Labour government to undertake 'serious studies' to consider improvements to Polaris. The work was carried on through four changes of government to completion in the early 1980s, with no public debate, and the first mention of the word Chevaline in Parliament was by Defence Secretary Pym in 1980. The overall estimated cost of the programme was by then over £1000 million, a large amount not to have gained even a passing mention in defence estimates.

Commons Select Committees

One way in which the House of Commons tries to exercise control over government actions and to obtain information is through the Select Committees. The Defence Committee is one of fourteen Select Committees appointed in 1979 to shadow individual government departments. Its task is to 'examine the administration, policy and expenditure' of the UK Ministry of Defence. Specific subjects for inquiry are not normally referred to it by the House, nor are the government's legislative proposals, which are generally examined by Standing Committees.

The committee has the power of subpoena on UK territory in respect of 'persons, papers and records'. This means in practice that the committee may call for evidence from and question ministers, civil servants and anyone who is involved in the subject in any way. While they may recommend changes in policy and expenditure, they do not have the power of appropriation to give effect to these recommendations – as does the US Senate Armed Services Committee for example – this would require substantial constitutional change. Ministers retain the right to withhold information 'in the national interest', and the committee may not have access to papers containing the advice which civil servants give to ministers.

The Defence Committee consists of eleven backbench MPs drawn from all parties and a small full-time staff (who are officers of the House of Commons and not civil servants) supplemented by part-time specialist advisers.

This committee represents the furthest limit to which inquiry can be made on defence issues by the representatives of the people of Britain. For example, it began its inquiry into Strategic Nuclear Weapons Policy on 25 June 1980, and completed its hearings on 7 April 1981. The Committee Minority Report was frank in expressing its frustration that the government announced its decision to buy the US Trident system on 15 July 1980, seven months before the Defence Committee published its first evidence, and eight months before the House of Commons debated the issue. They stated that:

> 8. From the preceding paragraphs it will be clearly seen that Parliament's role in the decision to procure a successor system to Polaris has been limited to endorsing a decision already taken. Decisions on defence, and on Britain's strategic nuclear deterrent have historically been taken by a small elite of very senior Cabinet ministers, civil servants and service chiefs, and this present decision was certainly no exception.
>
> 10. The Government came to the House and invited it to endorse the Trident decision when the Committee was still deliberating. We saw no reason for action by the House before the Committee reported, and consider the Government's actions in this respect to be less than courteous to both the House and its Committee.[6]

As with any interrogation or dialogue, the success of Select Committees depends on the ability of MPs to ask the right questions and the willingness of civil servants to answer them: neither of which can be absolutely relied upon. MPs often lack really powerful briefing, and fail to push an issue to its conclusion. Civil servants are determined not to get drawn into discussing policy questions or 'who gave what advice to whom' on any particular issue. When they appear before these committees, civil servants are under instruction to observe a code of conduct, laid down by the Civil Service Department, which regulates what – and what not – to say. These regulations, which extend to more than sixty paragraphs of restrictions, are known as the 'Osmotherly Rules' after the assistant secretary who drafted them on the instructions of Lord Bancroft, then Head of the Home Civil Service.

The Defence Committee regularly publishes reports of its inquiries, but it should be noted that a substantial proportion of its hearings are in camera and the reports are dotted with deletions of what is considered to be sensitive evidence.

The Public Accounts Committee (PAC)

This is the most powerful committee in the House of Commons consisting of fifteen backbench MPs, and chaired by a former cabinet minister from the opposition. It has been a public watchdog since 1862, keeping a check on the billions of pounds for which Permanent Secretaries (heads of departments) have to account. It is the only Select Committee to have a large specialist staff working directly for it: the Comptroller and Auditor General and his staff of 600 are independent auditors who examine the accounts of each government department to ensure that expenditure is in accordance with the law and Parliament's intentions.

In practice, the PAC has extended its review beyond this limited audit function to scrutinise government spending for efficiency and economy. This inevitably leads it into policy areas and political controversy, although the committee itself generally acts in a non-partisan way. Many PAC reports have received widespread publicity and have caused the government and administration embarrassment. Senior civil servants prepare meticulously for the grilling they receive.[7]

The Comptroller and Auditor General, however, failed to inform the PAC of the escalating costs of the Chevaline project, although he had access to the relevant papers at the time. The committee was reduced to publishing an indignant report[8] after the event. The committee, determined not to let the same thing happen with Trident, asked for and considered the Comptroller's report on Trident in 1984.

The Cabinet Office

> The Cabinet Office comprises the Secretariat, who support ministers collectively in the conduct of Cabinet business; the Management and Personnel Office (MPO), which is responsible for the management and organisation of the Civil Service and recruitment into it, training, efficiency, personnel management and senior appointments; the Central Statistical Office; and the historical section. Other functions are from time to time laid on the Office, some ephemerally and some permanently. Non-departmental ministers may be attached to the Office. The functions of the Cabinet Office (MPO) are in support of the Prime Minister in her capacity as Minister for the Civil Service, with responsibility for day-to-day supervision delegated to the Chancellor of the Duchy of Lancaster.[9]

The definition above is taken from the Civil Service Year Book, and it is quite difficult to discover more precise descriptions of what the detailed functions of the Cabinet Office are, exactly how big it is, and its internal structure. However, that it plays a very crucial role in the formulation of defence and foreign policy is without doubt. Many experienced observers regard it as the engine room of British central government.

The *co-ordination of policy* required by the increasing complexity of government business is achieved as we have seen through the Cabinet and its committees. It is a function of the Cabinet Office to provide the machinery for this collective, co-ordinating process: first, by advising on the working of the committee system and on the appropriate forum for particular topics, and where necessary on setting up new committees; second, by servicing the Cabinet and its committees. The Secretary of the Cabinet, his private office and senior members of his staff play a major part in the first advisory or steering function, and keep in the closest touch with ministers' private offices and with the permanent secretaries of departments. The second servicing function is the main business of the Secretariat from the Secretary downwards.

The civil servants in the higher ranks of the Cabinet Secretariat, with the exception of the Secretary of the Cabinet, are on secondment from other departments. In the case of Principals and Assistant Secretaries they come for two years, and Deputy Secretaries usually for three. This is a very deliberate policy for a number of reasons. Firstly, the Secretariat is organised according to functions – overseas, foreign affairs, defence, economics, home and social – and people closely in touch with those departments are needed. It is also thought that permanent Cabinet Office staff should not be built up because the Cabinet Office does not have policies of its own.[10]

Very differing views are held on this subject. Sir John (now Lord) Hunt, Secretary of the Cabinet, 1973–79, said in 1977:

'I think that I manage a small but very busy activity which sees to all the arrangements for the Cabinet and Cabinet committees – getting the papers round in time, the agendas, briefing the chairmen, recording the discussions, sending out minutes, and chasing up decisions. That is the main activity.'

Brian Sedgemore (MP) does not agree:

'Sir John did himself and his staff less than justice. The chairmen's briefs they draw up are powerful documents often leading in a certain and clear direction – often in the direction thought best by civil servants who have paralleled the committees which Sir John and his men service. . . . He is aided in this task by the fairly regular Wednesday morning meetings of the Permanent Secretaries which take place in Whitehall.'[11]

The Cabinet Office also used to house the Central Policy Review Staff – the 'Think Tank' set up in 1971 and first headed by Lord Rothschild, to give the Prime Minister advice independent of the Whitehall machine on strategic and policy matters. It was abolished by Margaret Thatcher, in a move eloquent of a reluctance to consider alternatives to established policy. By the standards of Whitehall, the think tank had a very small staff and consisted of little more than sixteen acute brains assembled from outside as well as inside Whitehall. Most of its reports were not published despite efforts by Members of Parliament to get the Prime Minister to give permission.

Above the suite once occupied by Lord Rothschild is the Joint Intelligence Organisation. It provides an early warning system for foreign and defence policy-making. Each week its current intelligence groups report to the Joint Intelligence Committee, which prepares a 'Red Book' of summaries which ministers receive on Thursdays.[12] This structure plays a key part in the combined intelligence gathering which produces the 'Threat Assessment' or estimate of enemy capability, which forms the basis for planning military needs. It is described later in detail under Ministry of Defence Intelligence.

The present Secretary of the Cabinet is also Head of the Home Civil Service. He chairs the Permanent Secretaries' Steering Group on Intelligence, and the official Committee on Security.

The Cabinet Office houses the Overseas and Defence Secretariat, which supports the Cabinet Standing Committee known as the Defence and Overseas Policy Committee. This Secretariat has the task of providing the briefs and background papers for the final formal stage in the decision-making process – the pinnacle of the political pyramid in defence decisions. Ministers in the British parliamentary system have such overwhelming timetables and such lack of support staff that it is impossible for them personally to go very deeply into the vital issues on which they must say yes or no. So they have to rely on the briefs provided by civil servants. By the time the papers reach this stage – the Overseas and Defence Secretariat – the proposals and options have been minutely filtered and refined.

In the case of a decision on a weapons system, the process has begun years before in the design and research laobratories, and the proposal has worked its way through the various committees of the Ministry of Defence to the point where the Secretary of State for Defence is in basic agreement with his own civil servants about the decisions to be taken, before he is prepared to defend it in Cabinet sub-committee. Considerable amounts of money may have been spent by this stage and the briefs, by the time they are presented by the Overseas and Defence Secretariat, will strongly recommend one particular course of action.

Although small compared with the MoD or Treasury, the Cabinet Office's influence on nuclear issues stems directly from the fact that it has the ear of the Prime Minister. It co-ordinates the actions of government departments on nuclear policy issues, and hence often frames the terms of internal Whitehall debates. The Permanent Secretary to the Thatcher Cabinet, Sir Robert Armstrong, widely acknowledged to be the most powerful civil servant in Britain, headed the team of civil servants that negotiated the Trident II (D5) deal with the US.[13]

THE CIVIL SERVICE AND DECISION-MAKING

The official task of the British civil service is to implement the decisions of government. The ministers in charge of major departments have the title of Secretary of State, and are in the Cabinet. Each minister is assisted by second rank ministers of state, and third rank parliamentary secretaries, forming a ministerial team. This team forms the total government representation in the ministry, and is temporary. The civil service is permanent and has its own system of grading posts according to the level of responsibility, status and seniority. Since the system is inherently hierarchical, these grades assume considerable importance.

Grade I *Permanent Secretaries* – the officials in charge of departments. In the case of the Ministry of Defence and the Foreign and Commonwealth Office, because their political heads are named Secretaries of State rather than ministers, they are known as Permanent-Under-Secretaries (PUS), and are responsible to the Secretary of State for the efficient operation of their deparment, including the accounting – seeing that the budget allocated to the department is spent in accordance with parliamentary decisions. Salary is approximately £43 000 p.a. and there were forty-two of them in 1984, all males.

Grade II *Deputy Secretaries* – are normally at the ratio of about three to one PUS. Salary is approximately £36 000 p.a., in 1984 there were 136 of them of whom five were female.

Grade III *Under-Secretaries* – sometimes called Superintending-Under-Secretaries, whose salaries have risen 18% on their July 1983 levels to £29 500 in November 1984. There were 505 of them in 1984, of whom twenty-three were female.

The Secretary of State (the minister) and the Permanent Secretary work closely together, but the minister needs a support service which the senior staff cannot provide. This is the Private Office, and it often becomes the gearbox between minister and department. This is the part of the machine where the struggle of wills between politician and civil servant is most obviously acted out, and the post of Private Secretary (filled by an able younger civil servant as a stepping stone to promotion) is the hot seat between Permanent Secretary and minister.

Most of the interaction between ministers and civil servants takes place in private, and is governed by the Official Secrets Act making it difficult to perceive what actually happens.

Power over the Civil Service

Effective power over the organisation of departments (for example, whether to abolish one and create others), and appointments of individual permanent secretaries lies with the Prime Minister, and the Head of the Civil Service.[14] Suggestions have been made that top civil service posts should be acknowledged as political appointments. The situation would then have more similarity with that in the United States where large numbers of posts in the executive departments change with each administration.

A British minister cannot appoint or dismiss the Permanent Secretary of his or her department. If a minister wishes to be rid of a Permanent Secretary with whom he or she cannot work, the issue goes to the Prime Minister via the Head of the Civil Service, and sometimes the Prime Minister would prefer to dispense with the minister.

In more general terms, civil servants work within formal bureaucracies which are designed to provide a framework for control. They are governed by rules and regulations and are responsible to their supervisors. Staff are expected to be non-partisan in their work and at more senior levels may not be publicly associated with any political party. However much they feel a government or minister is mistaken, having given their advice and having argued their case, if they cannot persuade the minister, they are obliged to accept the decision and implement it. It is the minister, not the official, who has the authority.

Power of the Civil Service

It is difficult to trace how civil servants exercise their power for several reasons. Firstly, their dealings are shrouded in privacy; secondly, politicans sometimes blame them for frustrating their policies, and civil servants almost never reply; thirdly, the whole ethos of the service is governed by the concept of *ministerial responsibility*, and civil servants are most reluctant to state publicly their role in any decision. There are however certain avenues through which they may exert very considerable influence.

Briefing Ministers

Before an election, party manifestoes are studied 'with great consuming care'[15] by the civil service. Detailed briefs are prepared for each possible incumbent, showing which part of a policy can be implemented and which cannot, and, according to Tony Benn and Richard Crossman, stating what the civil service policy is. Others would vehemently insist that there is no such thing as a policy of the service. The incoming minister is presented with the appropriate brief, which can and often does outline the direction that the department hopes the minister will follow, including information on projects which have reached a certain stage. Ministers, who may never have heard of these projects before, let alone have the knowledge to make judgements on their progress, are forced to rely on the 'expert advice' of those around them.

Permanence

Britain has general elections on average every four years, and it is unusual for ministers to stay in one post for that long. Hence politicians come and go, but senior civil servants normally spend many years working in one department. This makes it difficult for politicians to build up an expertise to match that of their staff, who also have a powerful sense of continuity in favour of persevering with existing policies.

Policy Framework

According to conventional constitutional theory it is the responsibility of ministers to decide both on the framework within which policies are decided, and the policies themselves, and the responsibility of officials to carry out the policies chosen. Lord Armstrong, who as Sir William Armstrong was head of the civil service, described a complete reversal of these roles.

> Obviously I had a great deal of influence. The biggest and most pervasive influence is in setting the framework within which the questions of policy are raised. We, while I was in the Treasury, had a framework of the economy which was basically neo-Keynesian. We set the questions which we asked ministers to decide arising out of that framework and it would have been enormously difficult for my minister to change the framework so to that extent we had great power. I don't think we used it maliciously or malignly. I think we chose that framework because we thought it the best one going.[16]

Control of Information

To establish the facts of any situation and make a decision, information is vital. Civil servants do have the power to be selective in the information they present to a minister, to restrict or delay it.

Mobilisation of Whitehall

A department can, if it wishes, use a whole range of internal pressures to influence a minister. A report of what the minister is proposing, spread to colleagues in other ministries, can stimulate a flow of correspondence from other ministries (usually drafted by civil servants) calling for interdepartmental committees to be set up, the preparation of detailed statistics, and similar delaying tactics.

Civil Servants and Accountability

The problem about any contact with government is the difficulty of finding out who exactly is in charge. Civil servants say they are not, in the end, to be held responsible for decisions. They merely carry out the will of their ministers. This pure model of the constitution is plainly unsatisfactory.

> It suggests that the working environment of the civil service is a kind of moral vacuum, where no sanctions exist and no named person is to be held accountable for anything he or she does. This is quite clearly nonsense.[17]

There are indeed several forms of accountability. There is financial accountability, whereby departments must account for expenditure in accordance with the law and parliament's intentions, exercised by the Public Accounts Committee of the Commons. There is process accountability, whereby departments must answer questions on the methods used to carry out decisions, exercised by Commons Select Committees. But civil servants appearing before these committees are under instruction to observe a code of conduct which decrees what they may and may not say. There is a definite limit to the information MPs can obtain in this way, and the documents which are presented are frequently re-written for the occasion.

The major aspect for discussion on defence nuclear issues is policy accountability. Civil servants refuse to be held accountable for policy, while ministers have by and large followed and extended an existing nuclear weapons policy which was laid down in the early 1960s. In some cases this has meant having to abandon the manifesto statements on which they were elected.[18]

The Civil Service and Defence Decisions

> The Minister is crucially a manager. The Permanent Secretary is a managing director.[19]

This reflection comes from a man who occupied the top civil service position in the Ministry of Defence for six years, serving under three different ministers. It may well be truer of defence than of other departments, for a number of reasons.

Weapons Development

The time necessary for a major weapons system to go from drawing board to deployment is now fiteen or twenty years: ministers usually spend less than four years in office. In fact, in a ten-year period from 1974–84, there have been no fewer than five Secretaries of State for Defence.[20] Enormous sums are committed at earlier stages in the development of a weapon. An incoming minister will therefore be presented with a series of programmes which are extremely difficult to stop.

Technological Expertise

This becomes more complex every year, and further out of the reach of the average minister. It seems clear from the civil servants themselves, in departments which deal with long-term and technologically very complex production, that they have a considerable influence over ministers. According to Sir Patrick Nairne, Deputy-Under-Secretary at the MoD:

> In the defence area, a minister has got to be able to stand up to the admirals and scientists. Even if he does his homework he'd find that very difficult. I do honestly think that the terms of trade, so to speak, have moved in favour of the official.[21]

THE MINISTRY OF DEFENCE

The department responsible for implementing government defence policy and for control and administration of the armed forces, the Ministry of Defence (MoD) has a total payroll of over half a million people, including 325 000 in the armed forces and 195 000 UK civilian staff.[22] The MoD comprises one third of the civil service. From its headquarters in Whitehall it controls research establishments, factories, training centres and bases all over Britain and in far-flung corners of the world.

The MoD is the largest single customer of British industry, with an equipment budget of over $9 billion in 1985–86. In employment terms expenditure on defence equipment sustains about 225 000 jobs directly and a further 180 000 indirectly.

This massive bureaucracy and its armed forces are under the control of a hierarchy of civil servants and of uniformed officers, and are responsible to the elected government through five ministers. The most senior of these ministers, the Secretary of State, reports to the Defence and Overseas Policy Committee of the Cabinet, while all five ministers are answerable to Parliament in traditional ways.

Evolution of the Present Structure

Until 1964 there had been four separate ministries – Admiralty, Air Ministry, War Office and MoD. Seven years after they had been somewhat painfully unified into the Ministry of Defence, the Rayner Report in 1971 recommended the setting up of the Procurement Executive to combine the responsibilities of weapons development, ordering, financing and monitoring into a single organisation headed by a chief executive. In June 1981 the present ministerial structure was established in an attempt to lessen rivalry between the services over budget allocation. The latest re-organisation, begun in 1984 by Michael Heseltine, reflects his passion for organisational structures, and a desire to manage defence resources more efficiently. The changes, outlined in the White Paper 'The Central Organisation for Defence', in July 1984[23] were implemented by early 1985, despite considerable opposition from service Chiefs of Staff.

Ministerial Structure

In June 1981 the present structure was established. In addition to the Secretary of State, the Prime Minister now also appoints two Ministers of State, one for Defence Procurement and one for the Armed Forces, each supported by a Parliamentary Under-Secretary of State. The responsibilities of the Minister for Defence Procurement include contract matters, research and development (R & D) establishments and nuclear manufacturing, while those of the Minister for the Armed Forces include intelligence matters, NATO, mobilisation strategy and weapons deployment.

Decision-Making in the Ministry

As with any large bureaucracy anywhere, it is impossible to locate within the Ministry precise points at which decisions are taken. Decisions are not taken, they are shaped, as we shall see in detail in the description of the procurement process.

In deciding what weapons systems to develop and produce, the Ministry has to weigh and balance many internal and external pressures: from allies within NATO, from competing service demands, from the defence industries' lobby, from party policy, and from intelligence reports of 'enemy' strengths. Crucial to the shaping of decisions are two phenomena: the preparing of briefs and the meetings of committees. Right up to the stage when the Minister has to make a formal decision on a weapon, thousands of pages of briefs are amended, refined and distilled – and the final brief which reaches his desk will be set out in such a way that one particular option is recommended. All the way through, this forum is punctuated by the meetings of committees, of which a few concern us particularly.[24]

The *Equipment Policy Committee*, chaired by the Chief Scientific Adviser, is concerned with all proposals for future weapons when the budget will exceed £25 million for development or £50 million for production. The committee decides whether or not a weapon will proceed to production. There is some doubt as to whether this committee discusses nuclear warhead issues, but it is certain that they are concerned with decisions on delivery systems which are nuclear-capable. They will often have to reach a compromise between service demands for new weapons acquisitions. Members of the committee include the Vice-Chief of Defence Staff, the Deputy Under-Secretaries for Defence Procurement and Resources and Programmes in the Office of Management and Budget, the Deputy Chiefs of Defence Staff for Programmes and Personnel and Systems, the Deputy Chief Scientific Adviser, Controllers of the Navy and of Aircraft, Master General of the Ordnance, the Head of Defence Sales and the Controller of R & D Establishments, Research and Nuclear (CERN).

The *Defence Research Committee*, also chaired by the Chief Scientific Adviser, sets broad objectives for long-term research and considers scientific and technical matters affecting defence policy generally. Its members are the same as the Equipment Policy committee, with the exception of the Vice-Chief of Defence Staff and the Head of Defence Sales.

From 1960–71, the former Defence Research Policy Committee, whose job it was to decide priorities over the whole field of defence, never had nuclear matters on its agenda. At the technical level the subject was dealt with separately within a small nuclear enclave, (over which the CSA also presided). According to Sir Ronald Mason that is still the way it is.

The Financial Planning and Management Group was created in 1977. Chaired by the Permanent Under-Secretary of State, its brief is to match the defence programme to the available resources. In financial terms this is where each service will finally accept its slice of the defence cake. Long-term defence costs, financial restraints and strategic requirements are balanced here. The committee reports to the Secretary of State for Defence, and brings together the eight most powerful staff in the MoD: the Permanent and Second Permanent Under-Secretaries of State, the Chief and Vice-Chief of Defence Staff, the Chiefs of Naval Staff, of General Staff and of Air Staff and the Chief of Defence Procurement. There is a joint secretariat provided by the Office of Management and Budget and the Secretary, Chiefs of Staff Committee, underlining the formal links between the three.

The *Chiefs of staff Committee*, chaired by the Chief of Defence Staff, enables him to consult the head of each service and the Vice-Chief of Defence Staff on advice to be given to the Secretary of State on strategy, military operations and the military aspects of defence policy (as opposed to financial, technical, scientific, etc). It may also consider matters submitted to it by its two sub-committees (the Principal Personnel Officer's Committee and the Principal Administrative Committee), and major resource allocation and equipment issues.

Structure of the MoD

This structure is of such labyrinthine complexity that few of the people who work there understand it. An initial grasp of it can be attempted by working counter-clockwise round Figure 3.1, looking separately and in some detail at the various parts of the Ministry which have an involvement with nuclear weapons decision-making: the Defence Council, Single Services Headquarters Organisation, Defence Intelligence, the Defence Staff, the Department of the Permanent Under-Secretary, including the Office of Management and Budget, Defence Science, the Procurement Executive, the Procurement Process and the Research Establishments.

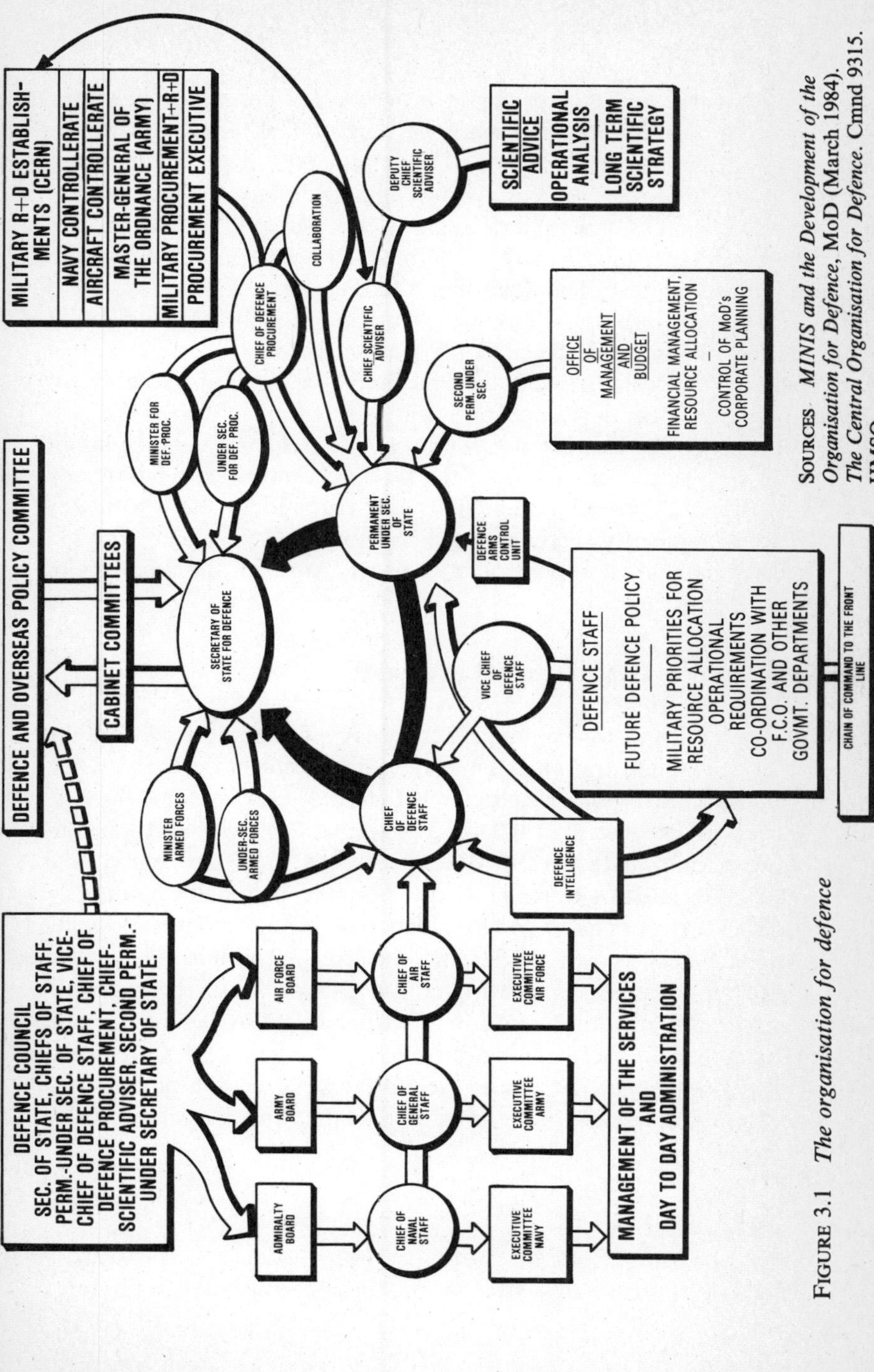

FIGURE 3.1 *The organisation for defence*

SOURCES *MINIS and the Development of the Organisation for Defence*, MoD (March 1984). *The Central Organisation for Defence*. Cmnd 9315. HMSO.

The Defence Council

The control of military power in a democracy presents a dilemma. Armed forces must not serve a political party, but at the same time they must be subject to civilian control. In the supervision and maintenance of the British Armed Forces, the mechanism designed to achieve this balance is the Defence Council and its Service Boards (the Admiralty, Army and Air Force Boards). The make-up of these bodies is both civilian and military. The Chairman is the Secretary of State (this helps to ensure the necessary link with Parliament): all his four junior ministers sit on the Defence Council, as well as the top civil servant from each major department, the Chief and Vice-Chief of Defence Staff and the Single Service Chiefs of Staff.

In wartime the Defence Council may meet frequently to take major decisions. The peacetime role of the Defence Council is to advise and assist in policy decisions, and it is not required to meet with any particular regularity. The number of meetings and the agenda depend on the personality and political style of its chairman, the Secretary of State for Defence.

Single Services Headquarters Organisation

This is responsible for fighting effectiveness, efficiency, morale and the day-to-day administration and management of each of the forces. Each Chief of Staff is responsible ultimately to the Secretary of State through the Chief of Defence Staff (CDS). The role of CDS has been substantially augmented in the Heseltine reorganisation. Since the Chiefs of the Army, Navy and Air Force no longer normally communicate directly with the Secretary of State, and formulation of military policy is no longer their responsibility, the CDS becomes the principal military adviser. He advises on strategy, on forward policy, and on which programmes should receive priority in resource allocation. He also has the ultimate responsibility for the planning, direction and conduct of all national military operations. His role is therefore crucial as one of the three most important people in the MoD and as the supreme military voice of the nation.

Defence Intelligence

Intelligence sources play a vital role in nuclear weapons decisions, not only in terms of preparation for use but, perhaps more importantly, in the very early decisions about what kind of weapons systems will be built.

The combined sources of intelligence reports put forward a 'Threat Assessment' to senior military and procurement staff. This indicates what the intelligence community estimate to be current and future dangers, covering indications of troop movements, as well as new weapons developments, tests, allocation of budget to new concepts, and so on.

> In the case of Britain's choice of strategic weapons, one of the major factors in opting for a submarine fleet was their present invulnerability, due to Soviet anti-submarine warefare techniques being estimated by Intelligence to be five to seven years behind the West. However, since the Trident II is expected to have an operational life well into the first decade of next century, the Commons Select Committee on Defence suggested that by that time a Soviet technological breakthrough could threaten the submarines' relative invulnerability.[25]

Some observers hold that 'threat assessment' plays a cardinal part in decisions on new weapons development; others take the view that these early decisions depend far more on technological innovation and economics, and that 'threat assessment' really acts more as justification. In either case, much depends on the interpretation given to raw intelligence data.

Data is gathered in the UK by four principal agencies, not all within the MoD:

The *Defence Intelligence Service* falls under the department of Defence Staff. The Director General of Intelligence reports directly to the Chief of Defence Staff. To him report a Deputy and Assistant Chief of Defence Staff and the Directors of specific intelligence units. The main objectives of this department, whose staff number between 800–1200, are to supply the 'threat assessment' to Ministry of Defence planners, and to supply information of tactical importance to the armed forces. The information is gathered from many sources including satellites, airborne warning and control systems (AWACS) on board aircraft, radar, remote sensors on land or the sea bed, surveillance from ships, aircraft and submarines. Allies, agents and defectors provide other sources of military intelligence.

The *Secret Intelligence Service (MI6)* with its headquarters in Century House, Westminster Bridge Road, SE1, reports to the Foreign Office on military intelligence and foreign counter-intelligence.

The *Government Communications Headquarters (GCHQ)* has superseded the Secret Service (MI6) as the UK's primary source of information for intelligence analysis, since the development of electronic technology. The Cheltenham headquarters, thought to have a payroll of 6500, monitors the airwaves, analysing and in some cases code-breaking, and then passing the relevant information to the intelligence network. GCHQ is part of the Foreign Office, but in essence is quite independent; answering only to the Permanent Secretaries Committee on Intelligence Services and the Joint Intelligence Committee which were mentioned before under the Cabinet Office. GCHQ has close links with the US National Security Agency (NSA) through a secret treaty known as UKUSA, signed in 1947. Between them these two agencies are capable of monitoring the entire globe.

The *Security Service (MI5)* with its headquarters at Curzon House, Curzon Street, W1, reports to the Home Office on domestic security and counter intelligence.

The interpretation and evaluation of data gathered is the responsibility of four main committees organised by a Co-ordinator of Intelligence and Security (a senior post in the Cabinet Office):

The *Joint Intelligence Committee* analyses intelligence gathered and examines short and long-term priorities.

The *Overseas Economic Intelligence Committee* carries out analyses of economic intelligence.

The *Official Committee on Security* supervises MI5, counter-espionage, domestic subversion, security, etc.

The *London Signal Intelligence Board* supervises GCHQ.

These four committees and the Co-ordinator of Intelligence and Security report to the top intelligence committee, which is the *Permanent Secretaries' Committee on Intelligence Services*.[26] The Permanent Secretary to the Cabinet is chairman of this committee, which passes on intelligence assessments of enemy threat, matters of national security and the intelligence community's own evaluation of the national and international defence situation to government ministers.

The Defence Staff

This is in effect a new department since the January 1985 reorganisation. Its main responsibilities are for future defence policy, and military priorities for resource allocation and operational requirements. It falls under the CDS working through his Vice-Chief with a new Unified Defence Staff, which is organised into four groups:

(1) *Commitments*, concerned with operations, exercises and logistics, a department handling NATO policy and operations, as well as the political and parliamentary aspects of military plans.
(2) *Systems*, sets the aims of military research, formulates operational needs, and deals with command, control, communication and information systems for the three services.
(3) *Strategy and Policy*, handles all longer-term defence policy and strategy and military nuclear policy work (with the exception of the day-to-day running of the strategic nuclear force, which is the responsibility of the Navy) including NATO, UK nuclear strategy, nuclear security, weapons, targeting and deployment. There are two departments, one handling military aspects of nuclear policies, the other handling the political, parliamentary and academic aspects of defence and nuclear policy.
(4) *Programmes and Personnel*, dealing with military priorities, resource allocation and service personnel.

It is clearly the Strategy and Policy area of the Unified Defence Staff which plays the most significant role in nuclear decisions, preparing the military position on long-term strategic needs, and with a responsibility to consider policy. The Unified Defence Staff will also of course have to weigh and balance competing service demands for particular weapons and allocations of the defence budget, but it is here that one of the comparatively rare areas (in nuclear decision-making) for long-term policy consideration exists.[27]

The Department of the Permanent Under-Secretary

The Permanent Under-Secretary's (PUS) Department controls the day-to-day running of the Ministry of Defence. It is headed by the PUS, who is permanent head of the department and Principal Accounting Officer, as well as being the channel for all civilian advice to the Secretary of State. It is one of the most prestigious and powerful posts in the British civil service.

Until January 1985 the PUS controlled the Defence Secretariats, bodies secret even by MoD standards. The Secretariats were organised into some twenty divisions each headed by an Assistant Secretary and identified solely by numbers. Defence Secretariat 12, for example, was responsible for NATO policy and strategy, overseas political questions, home defence and co-ordinating ministerial committee briefing, while Defence Secretariat 17 was responsible for national and NATO nuclear policy, nuclear and conventional arms control and disarmament questions. The Defence Secretariats were considered by some to be the heart of the MoD, working with and for committees and formulating policy options. They have now officially ceased to exist.[28] Their functions have been incorporated into newly-formed departments.

The Permanent Under-Secretary is one of the two principal advisers to the Secretary of State (the other is the Chief of Defence Staff). He is also responsible for supervising the accounts and long-term financial planning, acting as a link with other government departments. Most important of all, he is responsible for the co-ordination of the views of the procurement, scientific, research and financial staff into advice presented to elected ministers to enable them to arrive at decisions. The very permanence of the civil service gives it the great strength of continuity: ministers who come and go are obliged to make decisions on the options presented to them; the PUS is in charge of the options.[29]

Reporting to the PUS are the Chief Scientific Adviser, the Chiefs of Defence Procurement and Procurement Collaboration and the Second Permanent Under-Secretary who heads the new *Office of Management and Budget* (OMB).[30] This new Office, set up under the 1984 reorganisation of the MoD, concentrates and organises the PUS responsibilities for long-term financial planning, resource allocation and control of expenditure into four groupings:

(1) *Resources and programmes* – co-ordination of the MoD defence programme and its long-term costing. Major proposals are vetted here before going to the PUS.
(2) *Finance* – responsible for MoD financial management, cost control and parliamentary accountability, this department will develop 'Executive Responsibility Budgets'.

(3) *Personnel and logistics* – financial scrutiny of expenditure proposals on personnel and logistics.
(4) *Civilian management* – training and conditions for civilian staff, industrial relations with the MoD.

There is a clear relationship, according to the MoD, between the work of the Defence Staff, the OMB and the Financial Planning and Management Group. One of the key changes which has emerged is that a central budget will replace single service budgets. The Services are worried that they will have less say in how resources are allocated because there is no military voice in the OMB.

Inside the MoD there is concern that the OMB will become a 'treasury within'. It is also on the cards that the traditional role of the Accounting Officer will have to be re-thought.[31] This in itself implies possible changes in the relationship between the MoD and the Treasury.

The *Defence Arms Control Unit* was set up as part of the January 1985 reorganisation, and reports to the Secretary of State through the Permanent Secretary. It was formed out of the previous Defence Secretariat 17, a purely civilian department, and now has both civilian and military members, twelve in all. The head of the Unit is at Assistant Secretary grade. A new Policy Review Section is being added to the Unit, which considers itself a 'policy division', working out what aspects of a problem are major policy issues. The Unit liaises with the Foreign Office Arms Control and Disarmament Department, and the Defence Department; with NATO disarmament policy via the Special Consultative Group; and with the United States Department of Defence and State Department via joint working groups, as well as through the usual ambassadorial channels. It has no permanent presence at the Committee on Disarmament in Geneva, the only nuclear disarmament forum at which Britain is represented.

Defence Science

Given the extensive use of scientific and technological innovation for defence purposes, with nuclear and chemical weapons, crucial electronics components in delivery systems, and now space weapons technology, scientists are playing an important role as government advisers on defence. Their influence is extended through their status in the MoD Scientific Staff, MoD Research and Development Establishments, and in defence industries.

Within the MoD what appears to happen is that advice to ministers, research options and technological options within the defence budget are handled by the scientific staff, while responsibility for the actual management of research and development is vested in the Controllerate of R & D Establishments, Research and Nuclear (CERN) which is situated in the Procurement Executive. Into this structure are fed 'threat assessments' from military intelligence, and technical innovation for new weapons capabilities from industry.

MoD Scientific Staff

This is a distinct department within the MoD consisting of a Chief Scientific Adviser (CSA), who is appointed by the Prime Minister to advise the government on all defence scientific matters, a Deputy Chief Scientific Adviser, and four Assistant Chief Scientific Advisers. They are responsible for four key areas:

(1) *Projects* – to advise on new projects and changes in R & D programmes – (to give a view on the technical likelihood that a weapon will succeed).
(2) *Studies* – co-ordination of operational analysis (to advise whether, even if a weapon does succeed, it is the right solution).
(3) *Defence Operational Analysis Establishment* – (at West Byfleet) to advise on choices – they do modelling, which is using *quasi*-scientific methods for examining a weapons 'mix'; this includes war-gaming and the cost effectiveness of different possible solutions.
(4) *Nuclear* – to advise on the implications of defence nuclear policy.

Some observers feel that the importance of the Defence Scientific Staff has been maintained in the 1984 reorganisation of the MoD. The CSA now has a full-time deputy, and will be channelling advice from all three services to the Secretary of State via the Permanent Under-Secretary; previously each service had its own Chief Scientist, advising on its own weapons plans. However, the Commons Defence Committee had strong reservations about the changes, questioning whether the post of Deputy CSA was sufficiently highly graded, and underlining the fact that CSA's own position has been downgraded. We consider it important that this should not be permitted to make any difference to the status and influence of scientific advice within the Ministry, and regret that the Secretary of State does not regard CSA as one of "the range of distinguished people who must have access" to him.'[32]

Unlike ministers and their senior civil servants, the CSA has time to sit and think. He must advise on the long-term implications of policy, for instance on the 'Emerging Technologies' for post-1995 battlefield use and the political elements of the militarisation of outer space (since Britain is being asked to play a support role in current US plans). His influence can be extended in several ways, notably through the Defence Equipment Policy committee which he chairs. This is the MoD committee which finally advises ministers on any major weapons issue; being permanent, this committee is in a powerful position where weapons procurement programmes of fifteen to twenty years are discussed.

The Secretary of State is also advised by the *Defence Scientific Advisory Council.* The chairman of that Council prepares reports, which would cover nuclear weapons, and discusses them with the Secretary of State. The Council, which was reorganised in September 1980, surveys the full sweep of defence science and technology through a number of 'technology boards'. The members, each appointed in his personal capacity by the Secretary of State, all 'distinguished scientists in the academic or industrial world'. The names of members have not been disclosed by the MoD in answer to questions from MPs.[33]

The **Chief of Defence Equipment Collaboration** is a new post created in December 1984 for fostering international collaboration in defence technology, research and procurement with other friendly governments. This is symptomatic of attempts to share mounting R & D and production costs on major new projects, and an effort to improve NATO compatibility in weapons systems. Huge difficulties have been experienced with early efforts. Development of the multi-role combat aircraft Tornado, for instance, made in the UK, Germany and Italy, has been hampered by cost overruns, operational flaws and incompatibility of systems.

The Procurement Executive and the Procurement Process

Within the MoD, the Procurement Executive is responsible for research and development, production and purchase of all weapons systems (including nuclear systems), and all equipment for the Armed Forces.

The Chief of Defence Procurement, who reports to the PUS and who is charged with providing leadership and guidance to his controllers and his heads of department, is the senior civil servant accountable for all activities of the Procurement Executive. Under his direct control are its main sections:

(1) *The Controllerate of R & D Establishments Research and Nuclear* (CERN) is the Procurement Executive's military research and development department. Since the 1984 re-organisation and privatisation plans for the Royal Ordnance Factories, there are seven remaining R & D establishments which report to the Controller of CERN. Four of them are concerned with nuclear weapons or delivery systems and will be examined in detail later in this chapter.
(2) *The Systems Controllerates* which deal with ordering and purchase of all equipment for the services, headed by three systems controllers: the Controller of the Navy, the Controller of Aircraft and the Master General Ordnance (land systems). Each Controller can be used by any of the services; in other words, if the Navy needs an aircraft it will obtain it from the Controller of Aircraft. The Strategic Systems Executive, which handles the design and procurement of submarine-based nuclear weapons systems, falls under the Controller of the Navy.
(3) *The Defence Sales Organisation* has as its main task the promotion of overseas sales of the products of the Royal Ordnance Factories of which all but two are being privatised.
(4) *The Central Divisions* under the Deputy Under-Secretary of State (Defence Procurement) which are responsible for budgeting and procurement policy, including industrial and international aspects of procurement. They also direct purchasing policy and quality control arrangements.

The Procurement Process

The term 'procurement' covers the whole Ministry of Defence process of ordering a weapon from start to finish. The Procurement Executive is that major part of the MoD which initiates and conducts research and development on new weapons, checks progress, decides procedure, finances production and supervises tests.

As far as delivery systems and platforms (missiles, aircraft, submarines, helicopters, howitzers) are concerned, there is no clear dividing line between nuclear and conventional procurement. Many systems are 'dual-capable', fulfilling both a nuclear and conventional role.[34]

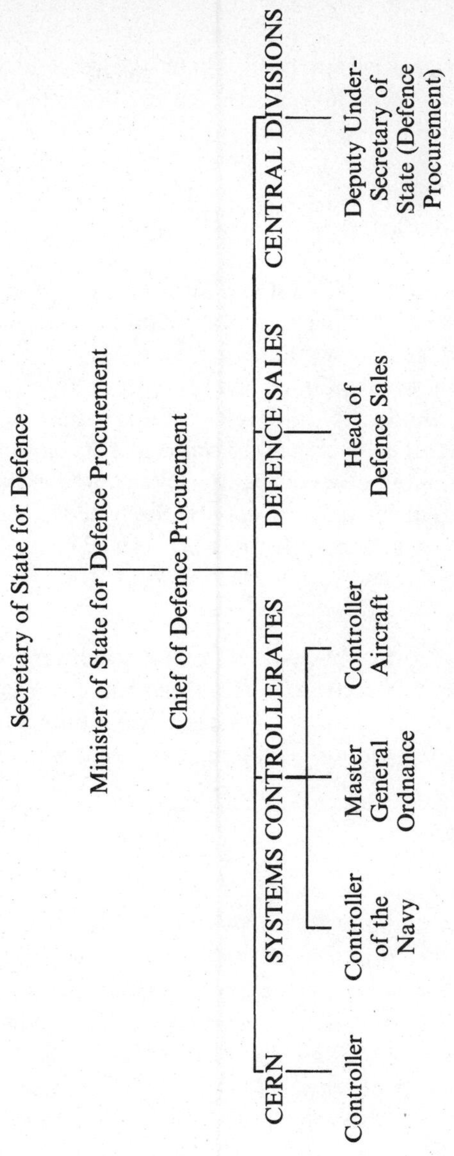

FIGURE 3.2 *Procurement executive top level organisation*

(For example, in the case of the UK's new multi-role combat aircraft, the Tornado, there are over 500 sub-contractors, each vital in their specialist contribution to the programme. Most of these companies would not think themselves involved in nuclear weapons work, despite the fact that one of Tornado's capabilities is to deliver nuclear bombs of up to one megaton yield.)

How the Process Works

The procurement process is complex and difficult to describe. What follows is an attempt to explain it in a palatable way, while adhering strictly to the facts as they are known.[35]

The idea for a new weapon, or the up-date of an old one, may be stimulated by a number of different forces within or outside the Procurement Executive: technical advances, new combat concepts, intelligence assessments of new enemy threats or the experience and information imparted by allies. It may be the outcome of the obsolescence of an old weapon, or arise from the perceived need always to keep ahead of opposing forces.

> 'Except in the short term', writes Lord Zuckerman, who was Chief Scientific Adviser to the MoD and the government for over a decade, 'so called "operational requirements" are always formulated around technological promises'.[36]

Staff Target

One of the Services submits a 'Staff Target' which outlines in broad terms the desired performance of the new weapon or piece of equipment. This Staff Target is drawn up by the Operational Requirements Staff within a Service and will have been prepared with the assistance of the appropriate Systems Controller (Land, Sea or Air), the research establishments concerned, and with some relevant sections of industry encouraged to put forward their proposals to meet the requirement.

Equipment Policy Committee

In July 1984 the Operation Requirements Committee and the Defence

Equipment Policy Committee were merged. This merger had been suggested for years, since the two committees' functions overlapped substantially and many of the members were on both committees anyway. One new committee, now known as the *Equipment Policy Committee*, replaces the previous two, supported by sub-committees dealing with individual areas of the equipment programme. It handles all proposals for equipment programmes having estimated costs exceeding £25 000 000 in R & D and £50 000 000 in production (the Tornado programme for example, includes 385 aircraft costing £19 million each). The Committee is chaired by the Chief Scientific Adviser and advises the Minister on which weapons items to include in the Procurement Programme. Its members include representatives from the Foreign and Commonwealth Office and from Industry, and expert advice is sought from outside specialists if required. The Equipment Policy Committee monitors the progress of a weapon through the procurement system, and must finally endorse it at at least three stages.

Feasibility Study

Once a Staff Target has been endorsed by the Equipment Policy Committee, and obtained ministerial approval if costs exceed £25 000 000 for R & D or £50 000 000 in production, it then passes to the Procurement Executive for a Feasibility Study carried out by the staff of the relevant service controllerate with the help of any establishment or industry sector from whom they need advice. The Feasibility Study is to decide whether the project is practical given the projected costs.

Staff Requirement

Following the feasibility study, the Service Department concerned, again consulting relevant Procurement Executive staff and industry, prepares a Staff Requirement which spells out in detail the required function and performance of the proposed weapon or piece of equipment. Again, it will be sent to the Equipment Policy Committee who will attempt to ensure that it coincides with strategic concepts, and that it is soundly based, technologically and financially. At this stage, the first major commitment to funds is made. If cleared for technical risk and threat assessment it is submitted to the *Equipment Policy Committee* for further endorsement.

Finally, the weapon goes for trial, contracts are placed, and production commences. This part of the process can be quite lengthy, with more trials at various stages before the weapon goes into service. Ministerial approval has to be given at each stage for projects where the costs exceed £25 000 000 for R & D or £50 000 000 in production.

Nuclear Warheads

In the case of Polaris and Trident, the missiles come from the US and are fitted with British warheads. All warheads are the concern of the Controller of R & D Establishments, Research and Nuclear (CERN) who controls the Atomic Weapons Research Establishment, AWRE, Aldermaston, and the warhead facilities at Burghfield and Cardiff. It is not known whether nuclear warhead production follows the process outlined above. However, in the case of the Chevaline update of the Polaris warhead, it is certainly doubtful. The whole project was shrouded in secrecy – conceived, designed and developed by the Aldermaston team.[37] It was never brought to the Defence Equipment Policy Committee, and only a very restricted number of Cabinet Ministers were aware of it at major decision points in the first seven years of its development.

In the case of nuclear weapons decisions, it seems that the Secretary of State for Defence is not obliged to get the full agreement of the Cabinet or even of the Defence and Overseas Policy Committee to the research and development stages. He can, in the words of a former Chief of Defence Procurement, 'do what he likes', except that he has to have the agreement of the Chancellor of the Exchequer and the Foreign Secretary. He can opt to restrict awareness of a project in this way for 'straight political reasons' (that other Ministers might disagree), not for reasons of security.

Military Research and Development

The Controllerate of R & D Establishments, Research and Nuclear (or CERN) is the *management* arm of nuclear pioneering for the MoD: it operates and manages all nuclear research and development (with the exception of the Defence Operational Analysis Establishment, mentioned under Defence Scientific Advice). This is carried out at the Atomic Weapons Research Establishment at Aldermaston, the Admiralty Research Establishment at Portsmouth, the Royal Aircraft Establishment at Farnborough and the Royal Signals and Radar Establishment at Malvern.

These various establishments also channel information on what is technically possible to the Secretary of State. The Controller of CERN thus sits on the important Equipment Policy Committee, and is seen by the Chief Scientific Adviser as his opposite number in managing research on nuclear questions.

The influence of the Official Secrets Act on defence issues in Britain has been the subject of much discussion, but its effect on the quality of scientific debate on the nuclear issue has received little attention. Scientists who work at each of the research establishments described below may not publish or receive academic criticism of their work.

The Atomic Weapons Research Establishment (AWRE) Aldermaston

This is located near Reading in Berkshire, and is the site of all major nuclear warheads research and development in Britain. The only other sites involved in manufacture are the Royal Ordnance Factory in Cardiff, where radioactive materials are machined, and the Royal Ordnance Factory at Burghfield, near Aldermaston, where the final assembly of most nuclear weapons takes place. Aldermaston also supplies the technical back-up necessary for maintaining the nuclear weapons stock-pile.

All nuclear R & D was moved to Aldermaston in the 1950s under the control of the United Kingdom Atomic Energy Authority. This control passed to the Ministry of Defence in 1973, where it remains. Some further changes in October 1983 led to the more direct organisation of Aldermaston from Whitehall, rather than from within Aldermaston itself. Many of Aldermaston's staff had been 'double-hatted', holding both Whitehall and AWRE jobs.

AWRE's Director now only reports directly to the Chief of Defence Procurement on matters of production programmes. For all R & D programmes he reports to the Controller of CERN. In an appendix to the minutes of evidence taken before the House of Commons Defence Committee 1980–81[38] the roles of AWRE are described as being:

(a) to develop nuclear warheads for future UK strategic and theatre weapons;
(b) to manufacture fissile components for UK nuclear warheads; and
(c) to provide support in certain areas of the UK defence non-nuclear programme.

Aldermaston is a major local employer, currently with 5000 staff. The running costs are not known since it is not government policy to reveal operating costs of individual defence R & D establishments.[39]

Structure. AWRE has five major departments with associated divisions (all supported by their own engineering, technical, health and safety and supply services).

The Warhead Development Department is by far the most important and is the cornerstone of the establishment. The demands of the warhead development team will dictate the type of work carried out in all other departments. As far as home-produced nuclear weapons in the UK arsenal are concerned (for example, a new warhead for Sea Harrier or Tornado), the type of weapons are dependent on the design capacity at Aldermaston and the production facilities at Burghfield and Cardiff. As far as American-supplied weapons are concerned (for example, Polaris and Trident), Aldermaston warhead designers must work closely with US missile teams. An important division of this department is Mechanical Engineering Services. Engineers maintain close links with AWRE workshops and Royal Ordnance Factories for technical and programme links, and they control nuclear testing (previously carried out in Australia, now entirely in the US).

> David Owen, Labour Foreign Secretary from 1977–79, discussing the arguments preventing a total test ban, said in 1985: 'The shelf life argument was a deliberate diversion, the protection of a vested interest by the nuclear testing laboratories at Los Alamos, Livermore, and Aldermaston. US and British nuclear scientists worked in cahoots, sensing that their job security and scientific satisfaction were threatened. They lobbied hard inside the military and political establishment. They were even prepared to undermine the politically agreed negotiating stance of their governments; in the UK the insubordination reached such a serious level that one official had to be disciplined.'[40]

The Weapon Electronics Division is responsible for R & D of weapons electronics, both nuclear and conventional. They also check the effects of nuclear radiation on electronic components. The Assembly Division includes in their work the study of conventional explosive triggers, and mathematical theoretical analysis of explosions. The Special Projects Division provides systems and project management on suitable research and development programmes and does feasibility studies on future weapons systems.

The Mathematical Physics Department studies and calculates the structural changes in materials and warhead mechanisms during explosions with the aid of computers. When a nuclear weapon explodes, temperatures and pressures far exceed those that can be produced in the laboratories, therefore much of the understanding of these explosions must be derived from theory alone. AWRE is the sole establishment in Britain with expertise in this field.

The Applied Physics Department develops the specialised optical-mechanical electronics equipment needed to measure and monitor nuclear explosions, with the US–Soviet Test Ban Treaty monitoring in mind, and evaluates lasers as weapons.

The Materials Department has four divisions: Chemistry, Chemical Technology, Explosives and Metallurgy. The Explosives Division researches high explosives for the Ministry of Defence generally, while the Metallurgy Division deals with plutonium, uranium and beryllium.

The Engineering Services Department designs and constructs new plants and buildings, carries out maintenance work, engineering operations (including power station operation); it has mechanical and electronic workshops, and provides machine tools and general engineering support and quality control.

In addition to these, there is an AWRE test department at Foulness which is licensed to fire the heaviest charges permissible in Britain to measure the explosive process. The measured explosions are used for Test Ban Treaty monitoring.

New Building. AWRE's A-1 Building, which manufactured the first British bomb, is being replaced by a new weapons manufacturing complex, the A90 Complex, at an estimated cost of £230 000 000. The building is being modelled on the Los Alamos Scientific Laboratory in the US. The decision was taken following a visit by the then director of Aldermaston to Los Alamos in 1978.

At the centre of the A90 Complex will be four large bays containing laboratories handling weapons-grade plutonium from Sellafield (Windscale). Plutonium 'billets' are machined into the required shapes for warheads and then sent by road to Burghfield for assembly. Of the four bays, two are to be used to manufacture Trident warheads. In another R & D into new generations of nuclear weapons will be carried out, while the last is for the recovery of plutonium from obsolete bombs. The building is expected to start production by 1986 and to reach full capacity in 1989.

AWRE appears to generate new British warheads over and above the 'irreducible minimum' of the submarine warheads. For example, the idea of nuclear depth charges for Lynx helicopters probably emanated from AWRE and was fed upwards. This would require considerable lobbying to justify the expenditure.

In the future, the structure of the new A90 plant with four bays will create a momentum for continuing development work on new warheads. The two bays for Trident manufacture will create considerable production capacity, which in turn will increase pressure to fund programmes to use that capacity.

Royal Ordnance Factories, Cardiff and Burghfield

All but two of the Royal Ordnance Factories (ROFs) are to be privatised. The two exceptions which receive little publicity and which do not appear on lists of ROFs in government consultative documents are those directly concerned with the manufacture of nuclear warheads. The Royal Ordnance Factory Cardiff, situated in Caerphilly Road, Llanishen, Cardiff, in the middle of a housing estate has the job of shaping and assembling depleted uranium (U235) and beryllium, both in metal form, into the precisely-shaped components needed for modern nuclear weapon 'triggers' and 'reflectors' used to maximise the efficiency of the explosive device. This also entails the recycling of old nuclear weapons and re-working of radioactive materials.[41]

The Royal Ordnance Factory Burghfield, situated near Reading in Berkshire, a few miles from Aldermaston, is where the final assembly of British nuclear warheads and bombs takes place. The factory, which covers some 300 acres and employs over 600 people, was built in 1941 and converted to atomic weapons production in the 1950s.

Day-to-day running of both factories rests with the Director, Atomic Weapons Factories. The existence of the two plants obviously depends upon Britain's continued commitment to a nuclear weapons policy, and, as with AWRE, employment depends on a continued flow of new work.

The Admiralty Research Establishment (ARE)

This is situated at Portsmouth and is the headquarters for all the UK's surface and underwater military R & D. It was formed on 1 September 1984 by combining the former Admiralty Surface Weapons Establishment, the Admiralty Underwater Weapons Establishment and the Admiralty Marine Technology Establishment.

Its responsibilities are to research and develop the UK's maritime warfare systems for the armed forces. This includes surface and underwater electronic warfare, sensors, communications, weapons systems for ships and submarines, command control and communications, anti-submarine warfare, torpedoes, mines and general research into all aspects of military marine technology.

The Royal Aircraft Establishment (RAE)

This is at Farnborough in Hampshire and is described by the MoD[42] as the largest research and development establishment in Europe; it has overall responsibility for the conduct and co-ordination of R & D for all aerospace activities, with the exception of engines, rocket motors and radar.

The functions of the establishment require it to maintain and develop expertise over an exceptionally wide range of disciplines: aerospace technology, avionics, navigation, sonar, telecommunications, weapons systems, chemical warfare, explosives, guided missiles, satellite communications and many more. RAE skills are used to support the Services, government agencies and industry, extending from basic research and the conceptual stages of aerospace projects, through development programmes for new operational techniques, as well as solving problems as they arise in service.

RAE is responsible for overall management and monitoring of all MoD funded research performed in industry and at universities. Relationships between RAE and the academic world are very close, with direct formal links existing with the Universities of Reading, Southampton and Surrey. RAE also emphasises the transfer of R & D results to industry, which further enhances the close links between the two.

On the international scene, RAE participates in the development programmes of multinational projects, such as Tornado, as well as in collaborative research, and acts as the technical agency for the exchange of information which takes place in agreements between governments.

It has extensive facilities for its scientific staff including large capacity digital computers, and a massive test plant with wind tunnels. A fleet of aircraft for R & D is maintained, and approach and landing systems are tested at RAE's main out-station in Bedford, with sites for ground and flight trials of weapons on the Welsh coast at Aberporth and Llanbedr, and on Salisbury Plain at Larkhill and at West Freugh in Scotland.

The main section responsible for nuclear-related developments is the Special Weapons Department. The designation of 'special weapons' is an indication of secrecy, but also of the fact that nuclear-related work is far from the central task of the RAE. Special Weapons Department had overall control of that part of the Chevaline programme undertaken by Farnborough, in conjunction with the prime contractor at the time, Hunting Engineering (which has a 'trials unit' at Farnborough). Control of the programme was later passed to British Aerospace for completion.

The Royal Signals and Radar Establishment (RSRE)

This is located in the Malvern Hills near Worcester and is the principal Ministry of Defence electronic warfare research and development establishment. The work at RSRE has far-reaching implications: electronics, especially advanced computer technology, is currently transforming military strategy.

Scientists at Malvern originated the concept of integrated circuits on silicon chips, and pioneered the development of liquid crystal display. They developed radar during the Second World War and are now working on electronics and infra-red technology. All these programmes – including work on a new thermal imager that can see through smoke or fog – might have important civil applications.

RSRE has an international reputation in the military world and ranks as one of Britain's major centres of scientific excellence. Former Malvern men, who quite like to be known as the 'Malvern Mafia', today occupy powerful positions – as Director of Aldermaston, Chief Scientist of the General Electric Company and as head of the 'Alvey Programme' – Britain's key research project into fifth generation computing.[43]

Like the Atomic Weapons Research Establishment at Aldermaston and the Royal Aircraft Establishment at Farnborough, it comes under the Controller, R & D Establishments Research and Nuclear (CERN) in the Ministry of Defence. It is mainly through this channel, and through contacts with the Chief Scientific Adviser, that technological advances emanating from Malvern will influence weapons decisions.

RSRE has two main departments:

(1) The Applied Physics Department is divided into the Physics Group, the Optics and Guided Weapons Group and the Electronic Devices Group. These three groups cover a range of research including the up-dating of radar, electronic counter measures, laser research, development of new materials for electronic application, infra-red research for night vision application and guided weapon research.
(2) The Systems Department is divided into the Airborne Radar Group, the Computing and Ground Radar Group and the Communications Group. The research of these three groups involves radar, guided weapon research, the evaluation of weapons systems, defence communication systems including satellites and the crucial computing and software research in which Malvern is acknowledged to be in the lead.[44]

> The military importance of electronics can be gauged by just one of the Navy's current weapons projects – the Type 23 frigate, now being built at Yarrow in Glasgow. This ship will have up to 200 micro-computers on board to control its weapon and sensor systems. The cost of the ship's electronics will be half the total cost of the entire ship's production.

The 1985 Statement on the Defence Estimates announced two interesting developments which indicate the increasing role played by private industry in design and development of new weapons:

(a) seven industrialists have been invited to participate in an advisory role in the MoD Headquarters Management Board for the research programme; and
(b) the MoD has assisted in the setting up of Defence Technology Enterprises Ltd, a private company whose object is to help identify potentially workable ideas generated in the research establishments.

Industrial Research and Development

The United Kingdom spends more on defence than on education or health. Defence is very big business: indeed of the total defence budget for fiscal year 1984–85, nearly half is spent on actual equipment. This clearly has considerable impact on the country politically and economically.

The defence industry uses its research skills to pursue the goal of corporate profit. If a company can develop a new product which fits a defence requirement, or, as very often happens, encourage a perceived defence need to fit a new technical concept, the rewards can be large.

Any company which wishes to remain among the top defence contractors must engage in substantial specialised research. A proportion of this is funded by the MoD (more than £150 million each year is spent with a variety of research establishments and companies), but most of the research is funded by the companies themselves, hoping to land an extremely valuable long-term contract.

In addition to the major participation of two giant corporations, British Shipbuilders and British Aerospace, the number of British companies involved in some aspect of manufacture of nuclear weapons delivery systems is very large. (There were, for instance, 800 British firms associated with the original Polaris programme.) A summary follows of those companies with the largest Ministry of Defence (nuclear) contracts, although it is often difficult to distinguish nuclear from conventional work, and the MoD does not provide a list of major contractors in order of priority.

British Aerospace (BAe), is the largest contractor to the MoD, which spends one-eighth of its total defence budget with BAe. The company functions through two main operating groups: BAe Aircraft Group designs, develops and produces a wide range of civil and military aircraft, including Jaguar, Tornado, Harrier, Hawk and AEW Nimrod. BAe Dynamics Group will supervise the installation of the Trident warheads and are well placed for Trident contracts. They manufacture a range of guided missiles which include Rapier and Sea Eagle as well as being prime contractors for all European space satellites. BAe Dynamics would be in the forefront of research and production for any European space weapon system.

British Shipbuilders is a large conglomerate of ship-building companies. It is organised into five divisions, the main one being the Warship Division including Barclay Curle plc, which makes missile launchers; and Vickers Shipbuilding and Engineering plc, which makes nuclear submarines (including Trident), warships, missile launchers and tanks. As partner of Rolls-Royce & Associates, they manufacture the nuclear submarine Pressurised Water Reactors (small nuclear reactors which drive the submarines).

British nuclear defence policy being based largely on submarines, British Shipbuilders have had strong influence with the MoD over the years in the building of Polaris submarines (of which they built two, while Cammell Laird built the other two). The Trident submarines are also to be built by British Shipbuilders.

Ferranti plc, specialists in military electronics, make a vast range of equipment from military computers to bomb fuses, and sub-contract for Polaris, Trident and Chevaline.

General Electric Company (GEC) in 1982 became the most valuable company on the London Stock Exchange with a market value of £5 180 000 000. It is one of Britain's major weapons manufacturers and the leading weapons electronics group, owning over a hundred companies including Marconi, Elliot Automation and the English Electric Company. GEC Marconi has supplied over half the UK's defence avionics over the last ten years, valued at £1 300 000 000. Subsidiaries, Marconi Space and Defence and Marconi Avionics, specialise in flight control data computers, defence satellite electronics, radar, and make the Polaris submarine torpedoes, the 7525 Spearfish nuclear-capable underwater missile torpedo, and have made parts for all of the UK's strategic nuclear systems.

Hunting Engineering plc was a major sub-contractor on Chevaline, (until replaced by BAe Dynamics), with heavy involvement in the nuclear weapons industry, making depth bombs and many other components, and is a major sub-contractor for Trident. Hunting collaborated with four Royal Ordnance Factories to produce the JP233 anti-airfield bomb (the biggest single British contract for munitions since the end of World War II, worth £500 000 000.)

Lucas Industries is a multi-national company of which Lucas Aerospace is the British subisidiary, manufacturing electro-mechanical components for military aircraft. 70% of Lucas Aerospace work is for the MoD.

M.L. Holdings are manufacturers of bomb release and handling equipment; also a sub-contractor for Chevaline, the Panavia Tornado (for British Aerospace) and the AV-8B Harrier (for McDonnell Douglas).

Plessey Co. are producers of electronic components for the weapons industry, and are in the top ten of MoD contractors, primarily for radio radar and field data gathering.

Racal Electronics plc. Through at least six subsidiaries (Racal-Automation, -Acoustics, -Communications, -Marine, -Microwave) Racal make a range of military electronic components, including UHF telecommunications, radar, marine radio and transmitters. Racal now own Decca, and are among the top ten MoD contractors with contracts worth over £100 million annually.

Rolls Royce are major manufacturers of aeroengines. Three-fifths of contracts in 1981 were military: Rolls Royce receive well over £1 000 000 000 from the MoD per annum. *Rolls Royce and Associates* was set up in 1959 to build Pressurised Water Reactors for submarines and is a consortium of engineering companies comprising Vickers, Foster Wheeler and Babcock, which design, develop and build the present PWRs.

Smiths Industries are manufacturers of avionics and flight control equipment, navigational systems and engine control systems for Harrier, Tornado and McDonnell Douglas AV-8B.

Thorn EMI a subsidiary of EMI Electronics, are manufacturers of weapon electronics, radar and a whole range of electronic warfare systems from bomb fuses to tape recorders.

Westland Aircraft plc are among the top ten MoD contractors, with contracts over £100 million. Military helicopter manufacturers, many of which are nuclear-capable, they have a heavy commitment to defence contracts, including a Marconi sub-contract for wing deployment on the new nuclear capable torpedo, bomb release for Tornado, and so on.

Influence of British Defence Industry on Weapons Decisions

From 1980 to 1984, 1400 senior MoD officials applied for permission to take jobs in industry, more than half of them in the defence industry. The phenomenon of ministry officials retiring from their posts to take up positions in industry has come under more public scrutiny since the Commons all-party Treasury and Civil Service Committee suggested the need for further safeguards.

The keenest 'poacher' of MoD talent is GEC Marconi, which, according to one former MoD Under-Secretary, himself a successful aerospace industry consultant, 'has more than any other company latched on to the value of ex-civil servants and ex-servicemen'. Marconi employs fifty former top MoD officials. Admiral Sir Richard Clayton was for three years

Controller of the Navy, responsible for the procurement of the Navy's weapons. In that position, he decided upon the development of the Sting Ray torpedo, whose estimated cost has soared in six years from £92 million to nearly £1 billion. In 1981, Sir Richard retired from the Navy and was immediately employed by Marconi's Underwater Establishment to head development and construction of Sting Ray. Clayton denies that while in the Navy he ever contemplated working for Marconi and shrugs off any suggested embarrassment for possible conflicts of interest. 'I'm proud of my association with the Sting Ray. I'm convinced it was the right decision for the country and the Navy.'[46]

The demand from industry for ex-ministry officials is easy to appreciate. Experienced civil servants who know everyone worth knowing in the higher echelons of government, offer valuable skills. They can help formulate corporate policy on lines most likely to be acceptable to those who advise ministers and they can recommend the most practical way of approaching a government department 'whether it's best to make a telephone call, send a 50-page memorandum, suggest a round table conference, or invite someone for a week end's shooting'.[47]

Clearly this situation is being exploited by the defence industry,[48] well aware of the advantage that inside knowledge of the MoD machine can give them over a rival company.

If career MoD civil servants see retirement to highly paid jobs in the defence industry as one of the perks of the job, this must raise questions as to the validity of some of their decisions while in office.

Although not yet on the same scale as in the United States, the same principle also works in reverse, where top industrialists take up senior ministry positions. The former general manager of Lucas Aerospace, for example, became head of the MoD Defence Sales Organisation; the former Chairman of United Scientific Holdings and Vice-Chairman of the Defence Manufacturers Association has been appointed Chief of Defence Procurement.

The informal power of major defence industrialists in Britain is much greater than their formal power, and is enhanced through several associations.

The Society of British Aerospace Companies Ltd is the trade association of Britain's aerospace industry. It has a powerful lobbying voice through the united presentation of aerospace interests. Using its influence, it sets up meetings with MoD defence procurement chiefs, the Defence Industries Council, the CBI and Trade Associations, to put its case for its share of the defence budget. The SBAC were invited by the House of Commons Committee, along with eleven individual member companies, to submit evidence on the current method of defence procurement.

The Defence Manufacturers Association was formed in 1976 to represent the interests of UK companies supplying products and services for the defence market throughout the world. It has a membership of over 300 companies. DMA works closely with the MoD, in particular with the Procurement Executive and the MoD Defence Sales Organisation. It also has close links with the British Overseas Trade Board and the Department of Trade and Industry.

The MoD Joint Review Board Advisory Committee consists of representatives from the SBAC, the CBI, the DMA, British Shipbuilders, the Electronics Engineering Association, and the Telecommunications Engineering and Manufacturing Association. The Committee represents the views of these associations to the independent Review Board Government Contracts. The Committee also advises the CBI in negotiations with the Treasury on recommendations made by the Review Board.

THE ROLE OF THE TREASURY

From its main building in Parliament Street, facing the Houses of Parliament, the Treasury pursues its goal of controlling public expenditure. Staff in the central Treasury number just over 1400, and of this total only 175 are in the senior administrative grades. The compact size and traditional key position of the Treasury has led to its senior posts being highly regarded in Whitehall.

In 1860 the Treasury in its present role was established as holder of the national purse, and structures were developed for departmental ministers to be required to make out cases for expenditure to Treasury officials.

Today, the Second Lord of the Treasury, the Chancellor of the Exchequer, is the effective Ministerial Head of the Treasury. He is responsible to Parliament for all Treasury business, including the overall management of the economy, the control of public expenditure and the direction of financial and monetary policy. He is also the Minister responsible for the two revenue departments, the Inland Revenue and the Customs and Excise, and for several smaller departments. The Chancellor is assisted by four junior ministers; the Chief Secretary, the Financial Secretary, a Minister of State and the Economic Secretary.

Essentially, the Treasury is concerned with the size of the overall defence budget including the cost of procuring and supporting nuclear systems. Treasury bureaucrats frown on financial surprises, and it is in the best interests of every department to engage in full and frank discussions with the Treasury before proceeding past the conceptual stage. The larger the size of the project, the higher up in the Treasury hierarchy these informal discussions go. Nuclear programmes are almost invariably discussed at the highest level.[49] In the course of these discussions, the general Treasury practices, of which other departments are so well aware, are: to let the spending department do the technical work; be sceptical; delay, probe, bargain and delay again; and look out for hidden expenditures – the thin edge of the financial wedge.[50]

However, the main section in the Treasury which deals with defence projects and expenditure, the Defence Policy and Materiel Group (DM), is small, with about twenty-four staff in total. It is under the supervision of two Assistant Secretaries and one Under Secretary, who reports directly to the Second Permanent Secretary. DM monitors the defence budget (£18.1 billion for 1985) and the 195 000 MoD civilian staff and 325 000 service manpower. The group was reorganised in August 1982 to bring control of defence expenditure and defence manpower together.

At ministerial level, DM normally deals with the Chief Secretary, who is also a member of the Cabinet. The Chancellor is also concerned on major issues, such as nuclear projects, which are then discussed by ministers collectively in Cabinet or Cabinet sub-committee.

Treasury Control of Defence Expenditure

The MoD negotiates with the Treasury a total amount of expenditure for all existing programmes. Within this total, it is up to the MoD to decide what it will spend on what. For major new projects there is a separate procedure: Treasury approval is needed for every project which will cost more than £12.5 million in development or £25 million in production.

The Defence Policy and Materiel Group of the Treasury is thus involved in all major defence expenditure decisions, dealing directly with the Office of Management and Budget in the MoD. It also has dealings with the finance staff responsible for each project. Its concern is that all options should be properly considered, with the object of achieving value for money. One of the main areas of argument is on non-competitive contracts, which account for a massive sixty per cent of the MoD's annual equipment spending. The Commons Public Accounts Committee has been particularly critical of the high profits enjoyed by defence industry, estimated to be £200 million between 1980–84. It should be pointed out that this practice of awarding contracts on a cost-plus basis, without competition, exposes the customer (in this case the MoD) to the risk of high cost overruns, encourages 're-designing as you go along' and sharply reduces contractor accountability.

> The Nimrod Airborne Early Warning project is an example of the dangers of cost-plus contracting. The final cost is estimated at £1 300 000 000, while at the beginning of the project (in the mid-1970s) the total cost was put at £300 000 000. The project is running two-and-a-half years late.[51]

With a total staff of twenty four, of whom only one is in the senior grades, the DM is ill-equipped to deal with the technological complexity of the arguments presented by the well-staffed departments of the MoD whom they are charged to control.

At times when the MoD is generously funded, as in the early 1980s when the defence budget grew by 3% per annum in line with government commitment to the overall NATO aim, the control task is even harder. The control systems are further weakened when the MoD asks for a 'separate budget allocation' as happened in the case of the Falklands. If Trident costs continue to rise very substantially, it is possible that the MoD might again ask for a separate budget allocation. The Treasury see no reason why Trident costs should not continue to be subject to normal MoD resource allocation procedures within a fixed defence budget.

In the context of nuclear weapons, the distinction between 'in-house' costs (those incurred at the MoD's own establishments) and 'extramural' costs should be remembered. In the case of Chevaline, all extramural commitments, 'from feasibility onwards, were the subject of formal Treasury approval, since the Department (the MoD) has no delegated powers for expenditure on nuclear weapons'.[52]

Within the MoD, the problem of long-term costings is a major headache. Supervision of these costings is carried out by the Financial Planning and Management Group, which is responsible for advising the Secretary of State on main resource allocation decisions. The system entails the annual costing of the forward defence programme for ten years ahead and involves the matching of present costing assumptions with long-term defence needs by a central complex of committees and officials under the direction of the Secretary of State.[53]

The Budgetary Cycle

This should be examined in the light of the UK budgetary cycle as a whole. Put briefly, each spending department costs its existing and continuing programmes, and may submit additional bills for new expenditure, to be considered in the Public Expenditure Survey. This part of the process culminates in a report submitted to ministers each June. The report is presented to Cabinet and forms the basis for what may become a protracted battle over resources.

The Cabinet takes decisions about the allocation of expenditure in the context of the government's overall economic and expenditure strategies, with the help of the Treasury's economic forecasts. The final decisions on the public expenditure planning totals for the years ahead – and on the allocation of resources to individual programmes – are taken by the Cabinet in the autumn. In recent years it has become usual for the plans for the next year ahead to be published promptly, in November or early December. Full plans for the next and two subsequent years are published in the *Public Expenditure White Paper*, nowadays in the early part of the calendar year. The main *Estimates* for the next year ahead – the formal request to Parliament for funds – are presented by the Treasury shortly before the beginning of the financial year, often on Budget day.

Parliamentary Control of Defence Expenditure

The Commons procedure for authorising public expenditure or 'supply' has become totally unsuited to the complexities of the issues and the process of decision-making. Parliament authorises the overall totals for public expenditure but does not analyse specific programmes in detail and cannot exercise advance control. It has more effect after the event, however, when it can carry out an examination of how the money was spent. The weakness of Parliament in this area has been a major problem in holding governments and civil servants accountable.[54]

Parliaments have, however, since 1982, introduced a number of changes to their financial procedures intended primarily to help them exercise more effective control over expenditure.

There is now a provision for three Estimates days every year, specifically allocated to debates and votes on Supply Estimates, including Revised and Supplementary Estimates. It is hoped that these will be used for real debates on expenditure in a way almost unknown since the 19th century.

The introduction of Estimates days, and the better opportunities they provide for debate of the Estimates on the floor of the House, seem likely to lead to an increased interest in Estimates by the fourteen departmental Select Committees set up in 1979 to shadow the major government departments. It is up to the Select Committees to decide the extent to which they scrutinise departmental estimates, and how to do so. But it is open to them if they want to call for further evidence from departments, to hold public hearings and to make recommendations to the House as a whole in the form of written reports. Their role, however, will remain advisory. They have no power to amend the Government's expenditure plans.

Defence Estimates therefore, are still voted each year, but only discussed (if at all) after decisions have been made on that basis. The Secretary of State presents to Parliament and publishes, in April each year the *Statement on the Defence Estimates*, in two A4-size booklets. Volume I deals with Defence Policy, the Management of Defence, Equipment, Procurement, Force Capabilities, the Services, and contains annexes which describe nuclear forces, defence industry, and so on, but in no great detail. Volume II is devoted to statistics, with sectors covering Finance and Trade, Service Personnel, Civilian Staff, Health Education and Accommodation, Defence Services and the civilian community. The Equipment sector does *not* break down into the costs of specific weapons systems, the so-called 'line items' which allow the US Congress to vet carefully what money is to be spent on particular systems *before* it is allocated.

FOREIGN POLICY, DEFENCE POLICY AND ARMS CONTROL

Arms control policy in Britain is engulfed within the broader process of national security policy formulation. In 1981 Lord Carrington, then Foreign Secretary, stated the official position:

> Arms control is part of our national security policy. Our arms control endeavours and our defence effort are two different but complementary ways of achieving the same end: peace with freedom to pursue our legitimate national interests around the world.[56]

By using the term 'complementary' the statement hints at the fact that arms control policy is directly linked, though a process of 'clearance', with defence policy. But the link is one-way. While arms control policy must be 'cleared' with the Ministry of Defence (MoD), defence policy need not be cleared for its arms control implications.

Arms control policy is the responsibility of the Foreign and Commonwealth Office (FCO). It is what is known as the 'lead' department for this issue which means that, formally, policy initiatives on arms control are generated within the FCO and that the Foreign Secretary or his junior minister responsible are answerable to Parliament. A separate arms control department, along the lines of the US Arms Control and Disarmament Agency, was considered and rejected by the Wilson government.

Policy Formulation

The formal channels for the formulation and execution of Britain's arms control policy are fairly clear (see Figure 3.3). Since the appointment of Selwyn Lloyd in 1951, an identifiable junior minister in the FCO has had responsibility for this aspect of foreign policy. At the official level a Superintending Under-Secretary is responsible (through the Permanent Secretary) for the two main departments relating to arms control. These are the FCO's Arms Control and Disarmament Department (ACDD) and Defence Department. The latter is described as being responsible for 'liaison with the Ministry of Defence on international aspects of defence' although in practice both departments liaise with the MoD on matters relating to arms control. The FCO Defence Department gained much greater importance in the arms

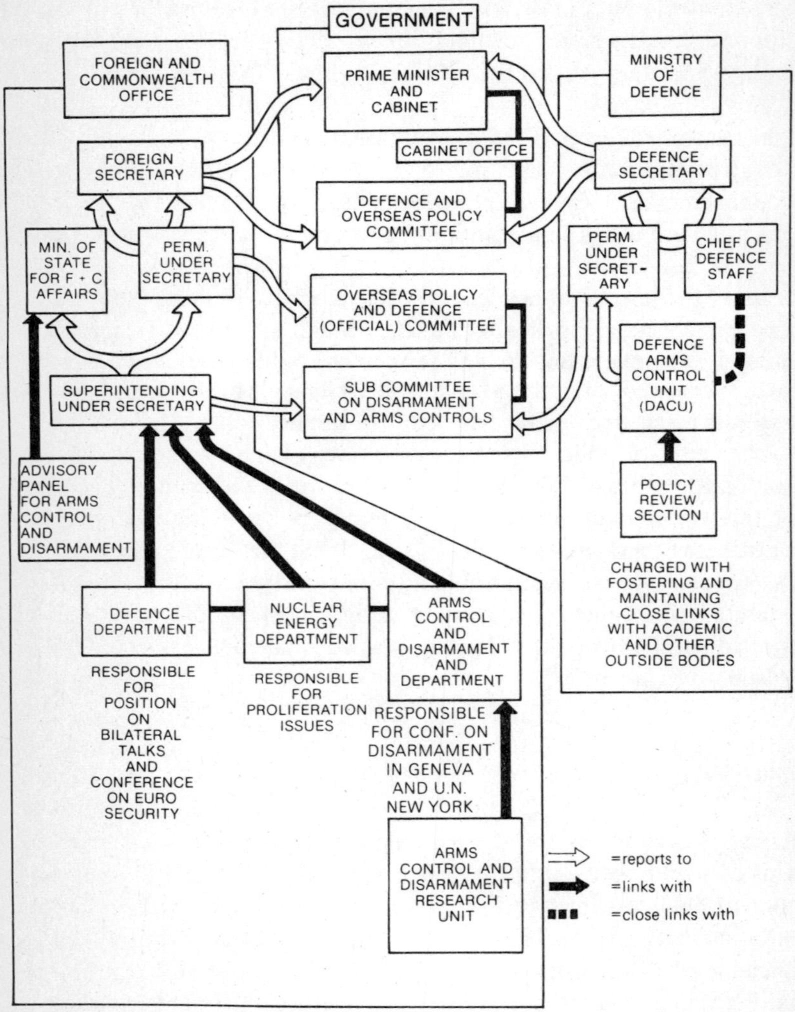

FIGURE 3.3 *Interdepartmental machinery for defence and arms control policy*

control field when the NATO section of the FCO Western European Department was transferred to its responsibility in 1979. This has meant that the most publicly visible arms control negotiations, bilateral talks between the USA and USSR fall within its remit. It is also responsible for the Mutual Balanced Forces Reduction talks and the Conference on Security and Co-operation in Europe. The longer-standing international fora for arms control negotiations, the Conference on Disarmament and the various United Nations bodies, are the responsibility of the ACDD. Within these departments responsibilities are variously divided up and given to particular desk officers, who deal with all the day-to-day running of their area, and instructions to the various British delegations are sent out.

An important 'triangle' for arms control policy exists between the Superintending Under-Secretary, the director of the Defence Department and the director of the ACDD, all civil servants. The quality of these three people and their working relationship is crucial to the direction and strength of arms control policy at the working level.

Clearance with the Ministry of Defence

Through the process of clearance, the MoD exerts considerable influence, verging on a veto, over FCO policy. The emphasis of its work is on protecting present and future weapons policy options.

The MoD has a Defence Arms Control Unit, consequent upon the reorganisation of the Ministry effective from January 1985. Formed from the part of Defence Secretariat 17 formerly responsible for arms control, it is now separate from the Defence Staff, and directly responsible to the Permanent Under-Secretary. The intention is to 'concentrate existing military and civilian arms control expertise within the Ministry' and 'include a policy review section charged with fostering and maintaining close links with academic and other outside bodies'. This sounds impressive, but the policy review section consists of two people, the entire Unit numbering no more than a dozen. It is not yet clear what role this Unit may play in the process of clearance.

Consensus and Consultation

Unlike American departments, the British civil service likes to form its policy through consensus. This is the formal function of clearance. It helps to avoid conflicts between departments and to settle differences sooner rather than later. Settling such differences is an important part of the work of committees of officials. They also try to ensure that the minimum workload necessary is sent up the chain of responsibility to Ministers. For example the Overseas Policy and Defence (Official), OPD(O), sub-committee on arms control brings together relevant heads of departments to settle issues before they are referred to the main committee of officials (see chart). The work of the full OPD(O), in turn, is to 'prepare the ministerial committee's business by preliminary discussion of the main issues and by eliminating, if it is possible, departmental disagreement on what is to be done'. Furthermore, 'frequently issues can be disposed of by this committee without waiting for the Cabinet (Defence and Overseas Policy Committee – DOPC) to be consulted'.[57]

This is not, however, to deny the significance of politicians in the formation of arms control policy, but rather to put it into a civil service perspective. The role of the politicans is to set the tone for policy discussions at the working level; to set the broad direction in which the 'ship' must be steered. Furthermore, through the clearance process between departments, officials will be acutely aware of the power structure between their own and other ministers.

The internal process of clearance in many ways mirrors the external process of 'consultation'. NATO and NATO allies will be consulted on policy changes and initiatives right from the working level. Summit meetings should be regarded as a final stage in a very long, detailed process of consultation or negotiation, depending on the circumstance. In international arms control fora, intense diplomatic efforts of consultation and preparation will precede any new initiative. Similarly, much of the policy process involves reacting to other countries' initiatives.

The Status of Arms Control

It would be wrong to assign the present sorry state of arms control negotiations, and with it the low ebb that Britain's arms control machinery has reached, to one government in particular. The concept of arms control has gone out of fashion with politicians in the West. It had its greatest popularity in the late 1960s and early 1970s when it was seen as the way of regulating the worst excesses of conventional arms developments and of producing the stability that nuclear deterrence strategy seemed to demand.

This fall from grace is reflected in many small but important changes in the FCOs arms control machinery. The early and exciting period when the Arms Control and Disarmament Research Unit (ACDRU) was a separate department bringing experienced academics, among others, into the Foreign Office machinery ended in the early 1970s and ACDRU was subsumed within the new ACDD with a smaller staff and reduced role. From 1976 the public information role became paramount as the research work withered. ACDD itself suffered cuts in staff. The UK Disarmament Delegation to Geneva had its Ambassador downgraded from first to third rank. These changes occurred at a time of heightened international tension. It is not surprising to find that new arms control initiatives have emanated from within the experienced staff of the Geneva delegation rather than from the depleted staff of the FCO itself, with their high turnover of staff leading to lack of experience in the complex and delicate field of arms control negotiations.[58]

THE 'SPECIAL RELATIONSHIP' WITH THE US

Ironically, it was the US refusal to continue atomic co-operation with Britain after the end of World War II which prompted the Attlee government to take the decision to build Britain's own first bomb in 1947.

> In October 1946, Attlee called a small cabinet sub-committee meeting to discuss building a gaseous diffusion plant to enrich uranium. The meeting was about to decide against it on grounds of cost, when Bevin arrived late and said 'We've got to have this thing. I don't mind for myself, but I don't want any other Foreign Secretary of this country to be talked at or to by the Secretary of State of the US as I have just been. . . . We've got to have this thing over here, whatever it costs.'[59]

Today, things are very different.

Britain has what is known as the 'Special Relationship' with the US – a remarkable web of political and strategic links in which nuclear affairs have been a key strand. Never before in history have two nations maintained such close and intimate ties in such sensitive areas as intelligence gathering and security policy.[60] From the British point of view, a British nuclear capability was one way of ensuring that British interests received adequate consideration in Washington and NATO discussions. At the same time, the US has profoundly influenced British nuclear policy through the transfer of data and components of nuclear weapons systems.

Weapons Supply and Collaboration

The Bilateral Agreement for Co-operation on the Uses of Atomic Energy for Mutual Defence Purposes was signed in 1958. Since that time all British warheads have been tested in the US; the agreement, and a further one the following year, enabled Britain to buy from the USA component parts of nuclear weapons systems, to receive information on the design and production of nuclear warheads and to exchange British plutonium for US enriched uranium. The 1958 agreements also produced a formal integration of targeting arrangements, with a British team working at the American targeting headquarters at Strategic Air Command in Omaha, Nebraska. The US also agreed to provide Britain's first nuclear submarine HMS Dreadnought with a nuclear propulsion system, for which purpose a new company, Rolls Royce and Associates, was established as the designated commercial organisation charged with handling the technology transfer. These agreements gave Britain privileges extended to no other nation, and ensured a much desired role in military discussions with the US.

In 1960 National Atomic Co-ordinating offices were set up in Washington and London, the British one in Washington forming part of the defence staff in the Embassy. (These offices continue to exchange atomic information and resolve any differences in interpretation of agreements.) The scene was set for the 1962 Nassau meeting when Macmillan persuaded Kennedy to let Britain have Polaris – the most advanced US missile force available, the British-built Blue Streak rocket having been cancelled in 1960 and the other replacement, the US Skybolt, also having been cancelled. The 1963 Polaris Sales Agreement was to set the seal on British nuclear policy for decades – a policy whereby its most important and expensive nuclear weapons have depended upon American designed and made missiles.

But weapons supply and collaboration is not a simple matter. For example, no British missile had employed large diameter solid fuel motors of the size and sophistication employed in the Polaris system; the Navy had little experience with advanced inertial navigation systems, which called for tightly controlled standards of care and accuracy. There were problems of secrecy. It has only recently come to light for instance that Royal Navy engineers have been prevented from carrying out detailed repairs on the BGR-15 sonar system because the US wished to safeguard the secrets of the system. Even when Britain is supplied with the necessary technical manuals, British repair shops remain under strict US quality control, and cannot carry out repairs unless certified by the US Navy.[61]

Hence complex liaison was necessary between the Special Projects Office (SPO) in the Pentagon (which managed Polaris), and the specially-formed Polaris Executive in the MoD. The organisational arrangements between SPO and the Polaris Executive were so effective that the Polaris system was delivered on time and with no cost overruns.

The next important nuclear weapons development (in the context of UK/US collaboration) was the British designed improvement to the Polaris system – the Chevaline project. As this differed significantly from the American MIRV systems, it could be argued that the UK was developing a vehicle which the US might find attractive. The move seems to have been designed to inject a degree of reciprocity into the Anglo-US relationship, and remove fears that the eight-year (1966–74) suspension of British nuclear testing would lead to a decline in the Joint Working Groups – the lynch pins of the relationship set up in the early 1960s. The procedures used to design the Chevaline warheads seem to have been the established ones, starting from an indigenous design and modifying it in the light of comments and advice from the US weapons laboratories, Lawrence Livermore and Los Alamos; implying extensive interchanges with the US Department of Energy as well as the Pentagon.[62] However the Chevaline project overran both in time and cost, in marked contrast to the Polaris purchase. This strongly influenced the next major British decision; to purchase Trident missiles from America.

The talks on Trident were handled principally by David Aaron, the number two official in President Carter's National Security Council, and by Robert Wade-Gery, a senior British Foreign Office official on secondment to the Cabinet Office. Their discussions were so secret that Aaron did not tell the American Ambassador in London, in whose residence he was staying[63] of his mission. It is thought that Carter's willingness to sell Trident to Britain may have been a *quid pro quo* for

Britain's willingness, agreed by Callaghan with Carter at the Guadeloupe Summit in 1979, to take cruise missiles as part of the Euro-missile decision. (See case-study in chapter six.) The decision in favour of buying Trident from America was taken by the Conservative government returned in May 1979, which later agreed to buy the more advanced Trident II, in order to maintain commonality with America. Britain agreed to man US Rapier air defence systems round US airforce bases in Britain throughout the life-time of the Trident programme and to maintain its naval deployments in the Indian ocean, in exchange for contributing only £116 million to the R & D cost of Trident II estimated at £9 billion.

In terms of Trident management, the successor to the Polaris Executive in the MoD appears to be the Strategic Systems Executive, a change of title which matches that of the Special Projects Office in the Pentagon to Strategic Systems Office. British firms are able to bid for sub-contracts for the Trident weapons system, but they must be approved by the US Navy before their bids are considered. US control is strict, with an elaborate system of offices and warehouses in Britain and America linked with the MoD offices in Bath, in London and in Washington.

In November 1984, the Mutual Defence Agreement of 1958 was updated and extended for ten more years.

Joint Targeting

The American Single Integrated Operational Plan (SIOP) is prepared by a sub-division of the US Joint Chiefs of Staff, the Joint Strategic Planning Target staff at Omaha in Nebraska – headquarters of American Strategic Air Command. A small British team works there, as part of a NATO team, and British nuclear forces receive their data on the British share of strike targets, identified according to RAF pilots as allied list numbers on a numeric targeting list. It is important to note that it was as part of the 1962 Nassau agreement that British nuclear forces were allocated in wartime to NATO Supreme Allied Commander Europe, although the agreement stipulates that the British government can withhold its forces if 'supreme national interests' dictate such a course.

US Bases in Britain

Over 28 000 US troops[64] are stationed at over one hundred bases in

Britain, some of them host to US nuclear weapons. This apparently falls under the Programme of Co-operation which the US normally signs with those countries where its nuclear weapons are stationed, but details of such a Programme have not been made available to the British public.

Intelligence Links

The intelligence relationship between the US and Britain has been close since 1947 when a secret treaty, known as UKUSA, linked GCHQ at Cheltenham with the embryonic National Security Agency in Fort Meade, Maryland.

The two agencies have an extraordinary joint capacity to intercept and decode signals world-wide, including military and diplomatic messages by radio, telex, teletype and microwave; all transatlantic phonecalls and satellite communications must be considered open to monitoring. Their computer systems can analyse four million characters a second.

GCHQ as already mentioned is formally part of the Foreign Office, but in effect is responsible to the Joint Intelligence Committee in the Cabinet Office, which began supplying the US with intelligence estimates as early as 1943.[65]

Over the years the relationship grew so close that GCHQ and NSA each stationed permanent staff in the other's agency, NSA moving into the UK in 1950. In the 1980s however, US concern over British security lapses reached a crisis with complaints that GCHQ had failed to provide full reports of Soviet spying at Cheltenham.

The CIA has been operating U2 and TR1 and SR71 surveillance aircraft out of Britain for some thirty years, and has been reported to fly an average five missions per week.[66] Britain also assists in the US satellite reconnaissance programme, providing numerous tracking installations – RAF Oakhanger, for example, was identified in Congressional testimony as a control facility for American satellites.[67]

BRITAIN'S ROLE IN NATO

Britain has been a member of NATO since the signing of the North Atlantic Treaty in 1949. The Alliance is based on the principle of collective security; while member states retain their full sovereignty, decisions are reached through the political, military and bureaucratic channels of NATO which deeply affect the nuclear policy of member nations. (See Chapter 6 on decision-making in NATO.)

One such area of NATO policy covers weapons. All NATO countries with nuclear roles and responsibilities need to reach decisions on which to buy, what characteristics they need, how many to deploy, which old ones to withdraw, where to increase research and development efforts. Many force development decisions are not subject to detailed political consultation in NATO, for example the US decision in 1975–77 to increase its F-111 fighter-bomber force in Europe. Since then, NATO nuclear force development decisions have tended to be politically prominent and collectively approved. The major force development decisions taken by NATO since 1977 are the following:

December 1979	Deployment of new US nuclear missiles, to begin December 1983.
December 1979	Withdrawal of 1000 obsolete short-range nuclear weapons within a year.
October 1983	Withdrawal of 1400 nuclear systems by 1988–9 (Montebello).
October 1983	Upgrading – in principle – of the remainder (about 4600) of NATO theatre nuclear forces (Montebello).

The NATO High-Level Group has been concerned between 1977 and 1984 with the structure of NATO nuclear forces; and its two major reports of 1979 and 1983 were force development reports. These reports formed the basis of the decisions listed above.

The NATO nuclear arsenal in the 1980s is becoming smaller; it is improving technically, and the centre of gravity is gradually moving away from old, 'dirty', battlefield nuclear weapons towards 'clean', hi-tech, conventional weapons, low-yield 'Mini-nukes' (small nuclear warheads) and accurate, reliable and effective longer-range nuclear weapons in the NATO–Europe theatre. This shift is the cumulative result of NATO's force development decisions of the late 1970s and early 1980s.

It is the High-Level Group which has framed the relevant force proposals and the Nuclear Planning Group which has approved them. The Defence Planning Committee and NATO Council have endorsed the sequence of decisions. The High-Level Group, a sub-committee of the Nuclear Planning Group, was set up in 1977 to prescribe nuclear policy requirements for the future. It is composed of senior personnel from national defence ministries, and chaired by the US Assistant Secretary of Defence for International Security Affairs. The British representative is the Second Permanent Under-Secretary at the MoD in charge of the new Office of Management and Budget. The activities of the High-Level Group are unpublicised and no communiqués of its meetings have ever appeared.

The Integrated Military Structure

In 1984 Britain devoted 95% of its total defence resources to NATO commitments. Britain belongs to NATO's integrated military structure, and most British armed forces are 'earmarked' for assignment to the 'operational command and control' of the Supreme Allied Commander in Europe (SACEUR) and the Supreme Allied Commander Atlantic (SACLANT) in circumstances of crisis or war.

In the nuclear weapons realm the British Navy, Army and Air Force all deploy nuclear forces committed to the NATO Alliance. Britain's fleet of four Polaris submarines is currently the core of Britain's 'independent nuclear deterrent'. They are deemed to be a 'contribution' to NATO's nuclear forces; they are targeted for use in wartime 'in accordance with Alliance policy and strategic concepts under plans made by SACEUR, save where Britain's supreme national interests otherwise require'.[68] This last phrase – the kernel of British 'independence' – means that British, not NATO, political authorities can have ultimate say over the use of Polaris and other British nuclear forces: for example, in a situation where the Americans and other NATO allies somehow desist from a European war which the British feel compelled to fight alone.

The British Army deploys entirely American-made nuclear missiles and nuclear-capable artillery with the British Army on the Rhine. The RAF and Fleet Air Arm of the Royal Navy deploy many squadrons of bombers and strike aircraft capable of delivering nuclear bombs and depth charges. These include the Buccaneer, Jaguar, Nimrod and Tornado aircraft assigned to either SACEUR or SACLANT. The Royal Navy also flies a variety of helicopters equipped with nuclear depth charges for attacking submarines. These are assigned to SACLANT, mainly for military missions in the East Atlantic.

British nuclear weapons decisions affect NATO as a whole, but there is no simple general explanation for the way these decisions are taken.

On an operational level decisions about deployment, targeting, missions etc, are clearly taken in close consultation with NATO military agencies: this is what the integrated military structure is all about. However, on fundamental questions about the composition of British nuclear forces, about their attendant rationales and doctrines, the British national policy-making apparatus is independent of NATO. In the course of the decision to replace Polaris with Trident, intensive US–UK joint deliberation occurred on many levels. This was predictable. But, even though the Trident decision is highly relevant to NATO's nuclear forces structure, and is indeed described as a 'contribution' to NATO nuclear deterrence, there was no high-level *NATO wide* deliberation prior to the British decision. This contrasts with the situation with the United States where decisions to deploy new American nuclear weapons on behalf of NATO have become subject to wider and wider NATO policy-framing consultation.

British nuclear independence used to cause inter-Alliance tensions in the 1960s. More recently, Continental Europe and the Americans have come to accommodate the distinctive British nuclear 'contribution'. This was formally acknowledged in the 1974 Ottawa Declaration of NATO Heads of State, a statement routinely referred to by British governments as authoritative NATO endorsement of British nuclear policy.

DECISION-MAKING ON THE CHEVALINE WARHEAD – A CASE STUDY

The Chevaline system was first operational in the summer of 1982 and is now deployed on all four of Britain's Polaris submarines. It consists of a new 'front end' for the Polaris missiles, and is designed to increase the probability of warhead penetration. To do this, Chevaline incorporates three new features which are intended to help overcome an Anti-Ballistic Missile (ABM) system such as the nuclear-armed Galosh which has been deployed around Moscow since the late 1960s.

The possibility of Soviet Ballistic Missile Defences nullifying Polaris was a source of concern in the late 1960s, and the Atomic Weapons Research Establishment at Aldermaston 'started looking at this seriously in 1967 and 1968 when the Soviet military deployment happened'.[69]

These initial studies led to the adoption of an American concept known as Antelope, which had been superseded in the USA because of the decision to develop fully MIRVed (Multiple Independently Targeted Re-entry Vehicle) systems. In 1970 a feasibility study was authorised, and this was followed by a project definition phase which cost £7.5 million.[70] The completion date of this project definition was the subject of an interesting disagreement between the MoD and the Commons Public Accounts Committee.[71] The MoD estimated that a five-year development programme followed by production would cost £175 million at autumn 1972 prices. Up to this time 'it was still a paper system, not much had been done in the way of actual hardware development'.[72]

Between 1972 and 1974 'the option to proceed with Chevaline development was kept open on the basis of interim funding for successive three and six month periods. Meanwhile, various other possibilities were considered'.[73] In fact the only realistic alternative was the American MIRVed Poseidon missile. According to Lawrence Freedman: 'During 1972 studies were conducted between the US Navy and the Royal Navy on the possibility of converting the British fleet to take the Poseidon SLBM. The studies resulted in a proposal to go ahead with a conversion'.[74]

However, when the initial cost estimate of £250 million was replaced by a more realistic one of £500 million, this compared unfavourably with the (current 1973) estimate of £255 million for Chevaline.[75] Also and probably more importantly, domestic political factors militated against the transfer of Poseidon missiles from the USA to the UK.

President Nixon was faced with a Congress which seemed intent on reasserting its control over defence issues, which was controlled by the opposition Democratic party, and which contained some Senators and Congressmen who were specifically opposed to MIRV technology and of the transfer of such technology to other nations.

Likewise in Britain the public purchase of the MIRVed Poseidon was not politically attractive. In particular it would have been likely to destroy the bipartisan consensus on nuclear weapons policy, and to rekindle public opposition to the nuclear deterrent.

The secret and therefore politically unembarrassing Chevaline project was thus chosen by the Heath government in January 1974 as the preferred system for the UK. It was to be deployed by the end of the decade. However, only £15 million funding for the next six months was provided, and then the general election of February 1974 returned Labour to power. Immediately after Labour's surprise win, a meeting of a Cabinet Sub-committee was called, including Wilson, Healey (Chancellor), Jenkins (Home Secretary), Callaghan (Foreign Secretary) and Mason (Defence). They decided to proceed with Chevaline, influenced undoubtedly by the fact that the UK's first underground test for nine years had been booked for May 1974. In April 1974 the full Cabinet was informed of a 'Polaris improvement programme' but chose to allow the wall of secrecy around it to remain.

Management of Chevaline

On completion of the 1974–75 defence review, a select group of Cabinet ministers decided in September 1975 to fund Chevaline to completion. Only then was a dedicated project management team set up, which 'immediately undertook a major review of the project, significantly increased the cost estimates, improved co-ordination, and established much tighter control over costs, time scales and production'.[76]

Subsequently, early in 1977, 'British Aerospace was appointed to co-ordinate the work of the various main contractors involved'.[77] Prior to these changes, from inception up 'until 1976 management of the Chevaline programme was largely in the hands of the nuclear scientists', whose performance was not outstanding.[78] Indeed, the House of Commons Committee of Public Accounts considered that 'the nature and extent of the changes found necessary in 1976 confirm that significant management and control weaknesses had existed for some time'.[79]

This lack of firm management control up until 1976 allowed the technological difficulties, and thus the cost of Chevaline to be persistently underestimated. As the Chairman of the British Aerospace Defence Group, Admiral Sir Ramond Lygo has stated, 'both HMG and British industry lacked the necessary technological and industrial base to undertake the Chevaline improvement programme without a long learning curve and expensive reliance on US firms'.[80]

Of the 250 development contracts for Chevaline, most were on a cost-plus basis. There was an almost threefold increase in the cost estimates between 1972 and the January 1977 estimate of £810 million (£495 million at 1972 prices).[81]

However, managerial inadequacies in coping with technical difficulties may partly be attributed to the political indecision which plagued the programme between 1970 and 1976.

This lack of firm political commitment reflected the fact that neither the Conservative nor Labour governments were totally convinced that Chevaline was the right choice. The Conservatives were keen to modernise (but were also tempted by Poseidon), whereas the Labour government was dubious of the strategic rationale which had led to the initiation of the Chevaline idea under the previous Labour government.

The Weak Strategic Rationale

Indeed, given the lack of convincing military rationale for Chevaline, once the signing of the ABM treaty had confirmed the superpowers' acceptance of the practical difficulties involved with ABMs, it is surprising that the programme continued at all. As David Owen has noted, 'Chevaline was not strictly necessary once the ABM treaty was signed on 26 May 1972.'[82] Certainly the strategic arguments for Chevaline were doubted by the 1974–79 Labour government, during which, according to Lawrence Freedman: 'Chevaline did not survive on its strategic merits'.[83] In David Owen's view, by the time termination was considered for the last time in 1977–78 'most of the Chevaline money had been spent or was already contracted for' and so it was continued, even though 'the only credible arguments for continuing were the political, not military aspects of our deterrence strategy'.[84]

The political arguments for continuing centred on the potential benefits of the technological expertise which had been acquired by Britain. Any future benefits from this (for example, in dealing with the USA) would have been foregone if the project had been terminated and the teams of specialists dispersed. *The Economist*'s view is that Chevaline was 'a weapon built not because it was needed but merely to keep the government's nuclear weapon research establishment at Aldermaston in business'.[85]

Chevaline may constitute a 'remarkable technical achievement',[86] but in retrospect does not appear to add significantly to the British deterrent, especially when the nature of Soviet ABM capabilities is considered. At its peak the Galosh system only comprised 64 launchers around Moscow, and that has since been reduced to 32. Moreover, although these have been modernised, they are still not technically convincing.[87] In the view of Air Vice-Marshal Stuart Menaul:

> Examination of the types of radar in the Galosh system and the state of Soviet computer technology, and even in the guidance systems employed in the interceptor missiles, would indicate that the efficiency of the Galosh system in shooting down ballistic missiles

would probably be less than twenty per cent.[88]

All in all, none of the arguments provides a convincing justification for spending over £1000 million (at 1980 prices) on a weapon system of little value to Britain. Chevaline was both excessively expensive and also of no significant military value, and as such appears to have been a failure on the part of the British defence decision-making system.

Control of Chevaline

As long as the government possesses a working majority in the House of Commons, Parliament will play no significant role in determining defence policy. Chevaline was not actually described to Parliament or even mentioned by name until January 1980 when Defence Secretary, Francis Pym, announced its existence. Dr John Gilbert, Minister of State 1976–79, was instructed not to refer to Chevaline by name at all. The only indications of Chevaline during the 1970s were bland references to Polaris improvement.[89]

No details of costs were provided, nor was there any requirement that they should be. As the Comptroller and Auditor General (C+AG), C. S. Downey, informed the Committee of Public Accounts,

> for defence projects it is the established and accepted practice – I believe on grounds of security – that the cost of even major projects are not shown separately either when expenditure is proposed or when it is being accounted for.[90]

Nevertheless, Chevaline was 'the largest matter in recent years not to have gained a passing mention in the annual defence estimates'.[91]

Since ministers rarely bring specialist knowledge of defence issues to their posts, and are hampered by the need to devote considerable time to political duties, they are inevitably dependent on the advice and briefings of their civil servants from the moment they take office. A major weapons project like Chevaline, 'would merit a special briefing for a Secretary of State in its own right' when 'the people in charge of this area would present the situation as it was to a new Secretary of State'.[92]

These nuclear weapons specialists within the MoD would, of course, be likely to support their colleagues working on nuclear projects at Aldermaston. Moreover, large, expensive and exciting projects such as

Chevaline are likely to engender considerable loyalty in those involved with them, and to thus build up their own bureaucratic momentum. Although the view that Chevaline was conceived in order 'to fill a gap in the work programme at Aldermaston', may be only one of many reasons for its birth, it is hard to believe that the investment in money and people did not prejudice decisions later on.[93] As Sir Frank Cooper admitted: 'Nobody to my knowledge likes giving up the job they are doing or the work they are doing . . . it would be idle to suppose that people do not have vested interests in programmes and things they are involved with.'[94]

Of course, such bureaucratic momentum is not exclusive to the MoD, and can be expected in all areas of government. Defence procurement differs, though, in that it is usually shrouded in secrecy – especially as regards nuclear systems – and so operates without the checks and controls that normal scrutiny may bring. In the case of Chevaline such secrecy may have allowed the commitment and enthusiasm of a few individuals to sway marginal decisions.

Formal decisions on Chevaline were taken in secret by a small number of Cabinet ministers – within a framework of continuity maintained by the bureaucratic momentum of career specialists in government departments. The crucial influence of these senior civil servants and the great secrecy surrounding nuclear projects not only exposes the lack of democratic control over these weapons developments and defence options, but also can lead to unjustifiable misallocation of resources – as in the case of Chevaline.

There was a burst of open government when Francis Pym, the Defence Secretary of the Conservative government returned in 1979, discovered that the £1 billion spent on Chevaline had been hidden from Parliament. He decided to reveal the project in the House of Commons in January 1980, to the fury of those on the Labour side who had been involved. 'It is one of the most outrageous, disgusting, most damaging examples of breaking the continuity of nuclear decision-making there has ever been', said one former Labour minister.'[95]

NOTES

1. What is known is largely due to the work of P. Hennessy, in his articles for *The Times* and in Cockerell, Hennessy and Walker, *Sources Close to the Prime Minister* (London: Macmillan 1984).
2. Mr Attlee's engine room was a monstrous 466 committees, accumulated in six-and-a-quarter years. Mr. Callaghan amassed about 190 in three years.

3. Parliamentary answer, 24 May 1979.
4. Hennessy and Walker, op. cit., p. 87.
5. Cambridge University Disarmament Seminar, *Defended to Death*, G. Prins (ed.) (London: Penguin, 1983).
6. House of Commons Fourth Report of the Defence Committee, Session 1980–81. *Strategic Nuclear Weapons Policy* (HMSO, 1981), p. xxxix.
7. H. Young and A. Sloman, *No Minister – an Enquiry into the Civil Service* (BBC, 1982) pp. 62–4.
8. House of Commons Ninth Report of the Committee of Public Accounts, *Chevaline Improvement to the Polaris Missile System* (HMSO, Mar. 1982).
9. Definition in *The Civil Service Year Book 1985* (HMSO, 1985).
10. House of Commons Expenditure Committee, Eleventh Report Session 1976–7, vol. II, part II (HMSO, Feb. 1977) p. 752.
11. B. Sedgemore, *The Secret Constitution* (London: Hodder & Stoughton, 1980).
12. *The Times* leader (28 May 1984).
13. C. Bowie and A. Platt, *British Nuclear Policy Making* (Rand Publications Services, Jan. 1984) R-3085-AF.
14. The tension between the Thatcher government and the civil service came to a head in Spring 1985 over the appointment of a non-civil servant as Chief of Defence Procurement at a salary of £75 000.
15. Sir Ian (now Lord) Bancroft, Head of the Home Civil Service on BBC Radio 4 (14 June 1981).
16. Quoted in *The Times* (15 Nov. 1976).
17. Young and Sloman, op. cit.
18. The Labour Cabinet sub-committee on Polaris replacement in 1978 clearly intended to proceed to a new generation of nuclear weapons in spite of the commitments made in the 1974 manifesto.
19. Sir Frank Cooper, former Department Secretary at the MoD, speaking on 'The Week in Politics', Channel 4, 13 July 1984.
20. 1974–76 Roy Mason; 1976–79 Fred Mulley; 1979–81 Francis Pym; 1981–83 John Nott; 1983– Michael Heseltine.
21. Sir Patrick Nairne, Deputy Under-Secretary at the MoD, 1970–73, quoted in Young and Sloman, op. cit.
22. *Statement of the Defence Estimates 1985*, vol. I, Cmnd 9430-I (HMSO, 1985).
23. Cmnd 9315 (HMSO, July 1984).
24. Information in this section is drawn from: Factual Amendments in Oxford Research Group Summaries, MoD, May 1985; *Civil Service Year Book* (1982, 1983 and 1984); House of Commons Defence Committee Second Report 1981–2. *MoD Organisation and Procurement*, vol. II (HMSO, 1982); R. Angus, *The Organisation of Defence Procurement and Production in the UK.* Aberdeen Studies in Defence Economics no. 13 (Dec. 1979); and L. Freedman, *Arms Production in the UK – Problems and Prospects*, Royal Institute of International Affairs (London, 1978).
25. *House of Commons Fourth Report of the Select Committee in Defence Session 1980–1* (HMSO, 1981).
26. D. Campbell, 'Inside the Sigint Empire' *New Statesman* (29 Oct. 1982); 'Unaccountable Empire Building', *New Statesman* (19 Nov. 1982) and 'Friends

and Others' *New Statesman* (26 nov. 1982).

27. Information in this section is drawn from: *Brassey's Defence Directory* (Pergamon Press) Published quarterly; B. Taylor, 'Coming of Age – MoD Evaluation 1974–82', *RUSI Journal*, Sept. 1983; and *The Central Organisation For Defence*, Cmnd 9315 (HMSO, July 1984).
28. Parliamentary written answer, Mr A. Butler, 29 Jan. 1985.
29. Sir F. Cooper, 'Perhaps Minister', RUSI Lecture, 6 Oct. 1982.
30. *The Central Organisation For Defence*, Cmnd 9315 (HMSO, July 1984).
31. 'Ministry of Defence Re-organisation: the Implementation of Change', a lecture given at RUSI on 28 Nov. 1984 by Sir C. Whitmore, KCB, CVO. Permanent Under-Secretary of State, MoD.
32. House of Commons Third Report from the Defence Committee 1983–84, *MoD Re-organisation*, Cmnd 584 (HMSO, Oct. 1984).
33. Written Parliamentary answer, 7 May 1985.
34. P. Rogers, *Guide to Nuclear Weapons* (Bradford School of Peace Studies, 1984).
35. Drawn from: R. Angus, op. cit. and L. Freedman, op. cit.
36. Lord Zuckerman 'Scientists, Bureaucrats and Ministers' in *Proceedings of the Royal Institution of Great Britain*, vol. 56. (1984) p. 223.
37. House of Commons Ninth Report from the Public Accounts Committee 1981–2, *MoD Chevaline Improvement to the Polaris Missile* (HMSO, 1982).
38. House of Commons Defence Committee Fourth Report 1980–1, *Strategic Nuclear Weapons Policy* (HMSO, 1981).
39. Statement to the House of Commons by Secretary of State for Defence Francis Pym Nov. 1980. However, the *Financial Times* quoted a figure of £100 000 000 p.a. in 1980.
40. In a speech given at the University of California at Berkeley, published in *ADIU Report*, vol. 7, no. 2 (Mar.–Apr. 1985).
41. *Royal Ordnance Factories*, multilingual brochure produced by the Procurement Executive, MoD.
42. *Defence Science*, Defence Open Government Document No. 82/89.
43. A. Milne, 'The Malvern Link', *The Listener* (10 May 1984).
44. *Defence Science 1982*, Defence Open Government Document No. 82/89.
45. House of Commons Second Report from the Defence Committee, *Ministry of Defence Organisation and Procurement*, 22-II (HMSO, June 1982).
46. *Observer* (8 July 1984).
47. *Observer* (19 Feb. 1984).
48. A brief survey conducted by the Oxford Research Group shows twenty-five senior MoD officials currently chairmen or directors of defence industries.
49. C. Bowie and A. Platt, op. cit.
50. H. Heclo and A. Wildavsky, *The Private Government of Public Money: Community and Policy Inside British Politics* (University of California Press, Berkeley, 1974) pp. 49–50.
51. *Jane's Defence Weekly* (2 Feb. 1985).
52. Memorandum from the Ministry of Defence, HC269, p. 2.
53. *MoD Organisation and Procurement*, House of Commons Second Report from the Defence Committee, vol. I, para 24 (HMSO, 1981).
54. *Financial Times* (12 Nov. 1984), points out:

'As an enquiry under the late Lord Armstrong reported, and the Treasury and Civil Service Committee of MPs has repeated, there is a clear case for more open discussion of proposed tax and spending measures, before it becomes effectively too late to alter them'.

55. *Economic Progress Report No. 155* (HM Treasury, Feb. 1983).
56. 'Arms Control and Disarmament No. 9' (Foreign & Commonwealth Office, Arms Control and Disarmament Unit, Aug. 1981) p. 21.
57. N. Simms, 'The Arms Control and Disarmament Policy Process in Britain', in H. Gunter Brauch, D. Clarke (eds), *Decision-Making for Arms Limitation: Assessments and Prospects* (Cambridge: Ballinger, 1983) p. 115.
58. Meanwhile the 'machinery' of both the Foreign & Commonwealth Office and the MoD is forced to fight rear-guard actions such as the one against Francis Pym's flirtation with chemical weapons re-armament when he was Defence Secretary. This led to the unheard-of event of the MoD issuing a statement that Pym's speech was not government policy.
59. Sir M. Perrin, who was present, *The Listener* (7 Oct. 1982).
60. Bowie and Platt, op. cit., p. 76.
61. *Observer* (19 Aug. 1984).
62. J. Simpson, *The Independent Nuclear State* (London: Macmillan, 1983).
63. *Sunday Times* (7 Apr. 1985).
64. The figures for March 1984 given by the US DoD Journal 'Defence '84' are 28 560.
65. D. Campbell, *The Unsinkable Aircraft Carrier* (London: Michael Joseph, 1984).
66. P. Malone, *The British Nuclear Deterrent* (New York: Croom Helm, 1984), p. 69.
67. P. Malone, op. cit., p. 72.
68. F. Pym (UK Defence Secretary), introduction, *Britain's Future Strategic Nuclear Deterrent Force*, Defence Open Government Document 80/23 (MoD, July 1980).
69. House of Commons Committee of Public Accounts Ninth Report Session 1981–2, *Ministry of Defence Chevaline Improvement to the Polaris Missile System*, HC269 (1982) q. 130.
70. Ibid., p. vi.
71. There was certainly confusion over when the crucial 'project definition' studies were completed. The MoD says they were completed in 1972 (HC269 – MoD Memorandum), whereas the Public Accounts Committee's criticism is that these studies were not complete even in 1975 (HC269. p. vi).
72. HC269, q. 180.
73. Ibid., MoD Memorandum para. 5.
74. L. Freedman, *Britain and Nuclear Weapons* (London: Macmillan 1980) p. 45.
75. HC269, MoD Memorandum, para. 5.
76. Ibid, pp. vi–vii.
77. Ibid., p. vii.
78. Ibid., p. vi.
79. Ibid., p. vii.
80. House of Commons Fourth Report from the Defence Committee Session

1980–1, *Strategic Nuclear Weapons Policy*. HC36, 1981, BAeD Memorandum, p. 183.

81. HC269, p. vi.
82. *Guardian* (1 June 1981).
83. L. Freedman, op. cit., p. 54.
84. *Guardian*, (1 June 1981).
85. *The Economist* (1–7 Sept. 1984).
86. P. Malone, op. cit., p. 21.
87. For a recent review of Galosh improvement see R. Hutchinson, 'CHEVALINE: UKs Response to Soviet ABM System', *Jane's Defence Weekly*, 15 Dec. 1984.
88. Letter to *The Times* (6 July 1981).
89. *Sunday Times* (10 Feb. 1980).
90. HC269, Appendix 1, p. 30.
91. L. Freedman, op. cit., p. 55.
92. HC269, qq. 252, 254.
93. Ibid., q. 177.
94. Ibid., qq. 162, 163.
95. Quoted in the *Sunday Times* (7 Apr. 1985).

4 France

In France, even more than in other countries, the formulation of policy and the technical development of weapons are closely intertwined in the nuclear decision-making process. The fundamental political orientations on military strategy and the ultimate decisions concerning nuclear weapons are taken by the President and a handful of advisers. Such decisions are, however, neither original nor made in a vacuum. They are influenced by the prevailing ideological climate and political circumstances as well as by the available technical options – and shaped by professional élites and by history. What is interesting, therefore, is the process which creates the options presented to the decision-makers – a process shrouded in secrecy and in which there is in France little or no public debate.

Four groups or communities control the whole decision-making process:

(1) the *political powers-that-be*, the President and a narrow circle of Ministers and experts meeting in the Conseil de Défense;
(2) the *nuclear community* dominated by the Commissariat à l'Energie Atomique (CEA) which controls the French nuclear fuel industry, and its military applications arm, the Direction des Applications Militaires (DAM) which develops and produces warheads and other nuclear explosive charges;
(3) the *defence community* which itself can be divided into three distinct parts for these purposes: the Ministry of Defence (administrators and strategy experts); the armed forces (especially the forces in operational control of the nuclear weapons such as the Air Strategic Forces) and the armaments community grouped within the Ministry of Defence's Délégation Générale pour l'Armement (DGA) which programmes, develops and controls manufacture of delivery systems such as missiles or bombers; and
(4) delivery systems *manufacturers* mainly in the fields of aerospace and aeronautics.

Within this framework, the decision-making process is multi-faceted, fluid and not clearly organised. On the one hand there are symbiotic relationships between the technical élites which control the bodies described above, and these reinforce the secrecy of the process. On the other hand, there are confrontations and differences between various factions, corporate interests and peer groups. The weight of history also adds to the complexity of decision-making.

The French nuclear deterrence concept – the 'Force de Frappe' – was developed and made into reality by General de Gaulle. Its foundations rest on the idea of French independence from the two super-powers, which also explains why the French armed forces and nuclear deterrent do not fall under NATO co-ordination, even though France remains a signatory of the North Atlantic Treaty. It was during the 1960s that the land-based strategic missiles and the nuclear submarines were developed and commissioned.

However, since the end of the 1960s the concept of 'dissuasion du faible au fort' (deterrence of the strong by the weak: an explicit reference to the different order of magnitude of the French arsenal from that of the super-powers) has gradually changed from a 'pure' deterrence theory to a nuclear war-fighting strategy incorporating the potential use of tactical nuclear weapons, (that is, theatre nuclear exchange capability based on anti-tank helicopter-launched nuclear rockets, nuclear mines and so on). The neutron bomb has also been developed. France has followed the same patterns as NATO nuclear forces, a pattern accelerated since 1981 under Mitterrand's presidency. In July 1985 the Socialist party modified its defence doctrine by asserting that the French nuclear detrrent should be used to defend West Germany as well as France.

The Power Elite in France

The decision-making process on nuclear weapons development falls within a narrow strata of the professional and administrative élite, and it is useful first to consider the way in which this élite is formed.

The French give great weight to academic qualifications. The right diploma from the right school is a pre-requisite for success in practically every field, at every level of qualification. There is a widely shared faith, even among those who make cynical comments about the system, that a competitive, hierarchical, uniform academic process is the best way to educate the nation's leadership and to determine who should rise, and how far.

The French élite is composed of people who attended one of the 'Grandes Ecoles' rather than one of the ordinary universities. Admission to these schools is based usually on the results of a single examination. In order to make a career in business, one attends a Grande Ecole de Commerce; for political administration there is the Ecole Nationale d'Administration (ENA); and for future leaders in various technical fields and in administration there is the most prestigious school of all, the Ecole Polytechnique. The ladder-climbing does not stop with admission to a Grande Ecole, particularly for 'les Enarques' (graduates of ENA) and 'les X' (the graduates of Polytechnique, so named after their school symbol – a pair of cannons crossed in an X). On graduating from these Grandes Ecoles the students are ranked according to academic standing and are then invited to join one of the 'Grandes Corps d'Etat', the membership of which opens doors (and is often the sole entry) to high level careers in particular areas. Membership in a Grand Corps generally involves a brief continuation of formal studies, and may involve participation in particular administrative or advisory functions for the State. The Corps that a graduate enters depends on the school he attended, and his class rank at graduation. The chart following explains the schools and their related Corps.

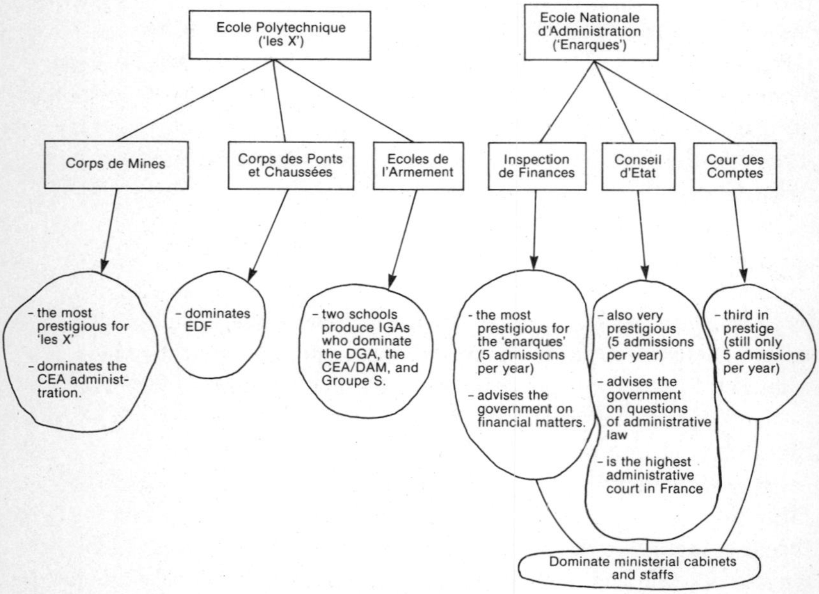

FIGURE 4.1 *The Grandes Ecoles*

This system of élites has several important effects. The graduates of any particular Grande Ecole form an informal network of co-operation and information flow. When an official needs information, advice, or help, he looks in the annual year book of the school which lists the positions and years of graduation of all its active graduates, to see if there is a schoolmate, or better still a classmate, who is an insider in the field in question who could help out.

In the field of nuclear arms development and manufacture, the system is far more coherent than any chart of linked organisations would imply. The simple fact is that more than 90% of the men in leading positions within this system are graduates of Polytechnique. Similarly, 80–85% of those occupying high level cabinet and staff positions in various ministries are graduates of ENA. Most of the administrators and engineers at CEA's Direction des Applications Militaires, and practically all of them at the Ministry of Defence's DGA are not only graduates of Polytechnique[1] but have gone on to become Ingénieurs Généraux de l'Armement (IGAs).

This phenomenon helps to explain some of the weaknesses of the armed forces within the defence community. Outside the Etats-Majors, most of the key positions are held by IGAs, not by soldiers – and high ranking officers are often bitter about this. However, the cohesiveness of this system of power élites is tempered by the often bitter rivalries that go on among graduates of different Grandes Ecoles and even among graduates of the same school belonging to different Corps. Competition between Electricité de France and the CEA (which erupted into a muted confrontation at the end of the 1960s) is considered as a battle between the 'Corps des Ponts et Chaussées' and the 'Corps des Mines' – although wider strategic and economic differences of interest were obvious.

Major institutions, and ministries as well, are often the territory of a particular Corps, and this territory is jealously guarded. Thus, not only are rivalries perpetuated, but the political leaders who appoint the top civil servants and professionals find themselves under considerable pressure to fill certain positions with men from particular Corps.

Cycles of Decision-Making in Nuclear Weapons Development

When a new nuclear weapon is developed, various options are generated at each different stage, eventually supported or rejected by the competing interests involved, and decisions are made to continue, discontinue or postpone each development option. From fundamental research with no apparent or immediate application to series-production, there are many physical stages of development. These include exploration of 'applicable' research, development and testing of prototypes and models, pre-series production, 'real-life' testing of first standardised products, tooling up and so on. At each stage, various decisions are taken, based on perceived needs at the time. These may be connected to a strategic purpose or to a perceived threat, or to the development of a similar idea elsewhere. They also reflect needs born out of the development process itself.

A number of cycles in the decision-making process in France can be defined. The affirmative decision to produce a weapon comes after a large number of active or passive decisions to continue research and development (R & D), and after much time and money has been consumed. Indeed the applied research stages may represent more than 30% of the final price of a weapon. France cannot afford to invest in the tremendous costs of development programmes which are to be abandoned later; it simply could not support the kind of vacillations that have been experienced over the B1 and MX programmes in the US, where a programme might reach the prototype stage only to be cancelled or postponed. Thus, in the realm of major weapons systems the decision to go for full production comes at an earlier stage in development.

This study therefore concentrates heavily on the early research and development stages. The main institutions involved (such as the CEA/DAM and the DGA) are described in the context of the process of decision-making in which they are most centrally involved. Figure 4.2 is a simplification of the complex liaisons involving the various interested bodies in each cycle.

One could begin from several points in the system. The choice here has been to start with the R & D policy formulation cycle, centred on the Ministry of Defence administration (Cycle 1); followed by the fundamental 'civilian' research (Cycle 2); the fundamental and exploratory military research centred on delivery systems and the production of those systems (Cycles 3 and 4); warhead research, development and production (Cycles 5 and 6) centred on the CEA; and lastly the ultimate as well as the primary source of decisions – the political and strategic decisions taken by the President and which epitomise the secrecy and ambiguities of the system (Cycle 7).

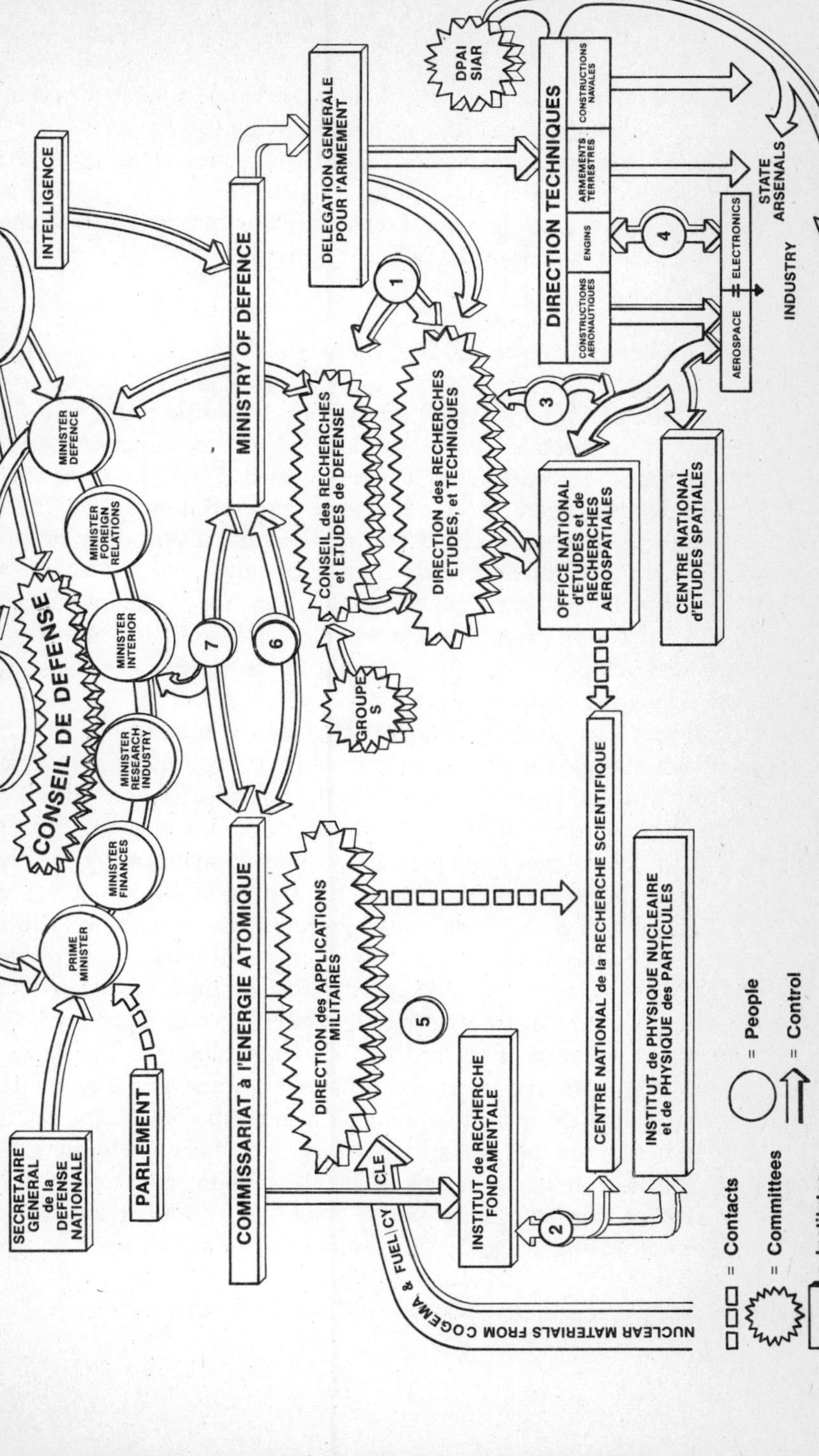

FIGURE 4.2 *Cycles of nuclear policy-making in France*

CYCLE 1 – MILITARY R & D POLICY FORMULATION

Nuclear weapons R & D efforts by all the organisations involved take place in the context of particular choices made (or supported) by the Minister of Defence as a result of R & D policy formulation exercises within the Ministry. The functions of the Ministry of Defence are therefore a good starting point, even if they do not control the ultimate decisions taken higher in the political process.

The Minister of Defence and His Ministry

These days the Minister is an entirely political position held by a close adviser to the President. The political side of the function has not always existed – at least not in its present form. Throughout the 1960s, French defence policy was the exclusive 'reserved domain' of Charles de Gaulle. At that time, either the Minister was a very close political ally and supporter of General de Gaulle,[2] or a highly placed technician; in any case he was expected to execute the orders issuing from the Elysée Palace. After de Gaulle resigned in 1969, the Ministry of Defence developed a life of its own; the Minister became a real voice within the government.

This trend has been slightly reversed under Mitterrand, who has surprised many with his 'neo-Gaullist' style. Although the current Defence Minister is no light-weight in the Socialist Party, the Ministry of Defence is again more of a 'transmission belt' for the strategic policies of a President more personally involved in defence issues than have been his two predecessors. Still, the Minister is part of the decision-making inner circle, a key voice on the Conseil de Défense, and is said to have the ear of the President at all times.

The Minister is advised and supported by a number of bodies within the Ministry, most of them with only a minor role in decision-making on nuclear weapons. First are two separate cabinets. The *Military Cabinet of the Minister*, a group of ranking officers, advise on military matters. Their role is fairly apolitical, a fact reflected by the relative permanence of the staff, some even surviving changes of Ministers. The *Civil Cabinet* on the other hand handles political matters for the Minister, at least the more visible issues. It is staffed by political appointees serving generally short tenures, who as as liaison with the outside world.

Another insider is the *Secrétaire d'Etat auprès du Ministre*, a position not thought to be of great influence, comparable to a junior Parliamentary Secretary in the UK. The Minister also has a personal adviser on scientific matters, the *Conseiller Scientifique*, one of his key links to the research community, who has a real say. He is a member of the important Conseil des Recherches et des Etudes de Defence (CRED). The current Conseiller is a nuclear physicist – one more link with the nuclear establishment.

Departments such as the *Secrétariat Général pour l'Administration* and the *Contrôle Général des Armées* are responsible for the general administration of the Ministry, respectively preparing and ensuring the execution of the Minister's decisions. The Ministry also contains the French foreign intelligence organisation, the *Direction Générale de la Sécurité Extérieure* (DGSE) whose role is discussed later.

Directly dependent on the Ministry is the *Direction des Centres d'Expérimentations Nucléaires* (DCEN), the military administration of the French nuclear test facilities which is responsible for carrying out test explosions in co-operation with the Department des Applications Militaires of the CEA. Despite its involvement in bomb development, it is not actually involved in the decision-making process – it merely provides military security and logistic support to the CEA engineers.

Other parts of the Ministry are far more central to the decision-making process on nuclear weapons development. In particular, the Délégation Générale pour l'Armement (DGA) is the embodiment of the arms procurement process, and its role will be discussed later. Two other bodies within the Ministry are central to R & D policy and strategic thinking: the *Conseil des Recherches et Etudes de Défence* (CRED) and the *Groupe de Planification et d'Etudes Stratégiques* (GROUPE S).

Long-Range Strategy Planning and R & D Orientation: Ministry of Defence Think-Tanks and Orientating Bodies

Of the two bodies above, the *Conseil des Recherches et Etudes de Défense* (CRED) is responsible for defining the overall orientation of military R & D. The CRED is presided by the Minister himself, and assembles the key members of the military planning apparatus: the armed forces, GROUPE S, the Secrétaire Générale pour l'Administration (SGA) whose role is rather passive, the Délégation Générale pour l'Armement (DGA) and the Minister along with the Scientific Adviser to the Minister and other *ad hoc* experts. This is a powerful council although it keeps a low profile. Each of its members, with the possible exception of the SGA, has a real voice and its decisions certainly affect the course of arms development. These people are continually receiving information about technical developments in the R & D community. They themselves are responsible for considering the evolution of defence doctrine and military 'need'. They discuss applications for new technologies and choose the projects which will be pursued far enough to become options for industrial development. Their discussions take place at a high level; lower down the line, the Direction des Recherches, Etudes et Techniques will also exercise this latter responsibility at a more detailed and practical level.

The other important body is the *Groupe de Planification et d'Etudes Stratégiques*, or GROUPE S, responsible for long range strategic planning. Known until 1982 as the Centre de Prospective et d'Evaluation (CPE), this think-tank and adviser to the Minister is composed of twenty-five members, mainly military engineers and representatives of the arms industry. It follows global military and strategic developments and the strategic consequences of new technologies and participates in the formulation of military doctrines and armaments policy.

> It was as the head of CPE during the 1960s that Lucien Poirier, then a colonel, came to be 'le père du nucléaire' as Charles de Gaulle's principal adviser on the classical doctrine of 'dissuasion du faible au fort', (the French doctrine of deterrence).

Although the relative influence of this body has declined since the 1960s, GROUPE S is still a crucial actor in French military development.

Officially, the role of GROUPE S places it between the politicians and Etats-Majors who define the broadest objectives of French defence policy and the departments responsible for weapons R & D, reconciling the needs of the first to the possibilities offered by the second. In fact, its role extends considerably beyond such a passive synthesis. At a time when the distinction between arms policy and defence policy is increasingly blurred – when technical possibilities cause strategic need rather than meet them – the ability of GROUPE S to affect the basic military thinking seems to be considerable. GROUPE S is also one of the strongest voices in the Conseil des Recherches et Etudes de Défence and as such has direct influence over the allocation of funds for R & D projects.

GROUPE S is also very well connected. In the nuclear field (and in the field of 'space age arms') GROUPE S is one of the military bodies most closely in touch with the secret developments at the heart of the Commissariat à l'Energie Atomique (CEA) and its Direction des Applications Militaires (DAM). In fact, GROUPE S is one of the sources of the policies which guide DAM's work (bearing in mind that it creates many of its own programmes and heavily influences the vision GROUPE S has of the nuclear future). Beyond CEA it can be assumed that GROUPE S is privy to the most secret information regarding enemy arms developments available.

GROUPE S is a key element of the technico-military complex. While it is meant to advise the Minister on behalf of the armed forces, it is always headed by an Ingénieur Général de l'Armement rather than by a soldier (thus, by a Polytechnicien who has risen through the DGA). So the Director of GROUPE S belongs more to the 'arms technology' community than to the Army. Traditionally, the Director of GROUPE S has been involved in nuclear arms development prior to working at GROUPE S.

In conclusion, the Ministry of Defence cannot be seen as a single entity in the nuclear weapons decision-making process. As a wing of the government, it extends over groups whose characteristics, tendencies and allegencies vary greatly. It includes soldiers, technicians, civil administrators and economists, scientists, intelligence experts and politicians. At its head is the Minister – a member of the political elite whose personal politics are less important than the views of his master, the President. While he can in theory override much of the will of the machine he is running, the extent to which the political and strategic options presented to him are shaped by this machine must not be under-estimated.

CYCLE 2 – FUNDAMENTAL 'CIVILIAN' RESEARCH

Seemingly far removed from the direct influence of Ministry of Defence strategies, the 'civilian' (non-military controlled) research community is often involved at the start of new weapons developments. The organisations responsible for nuclear weapons development make extensive use of civilian research facilities, especially in the fields of fundamental research. These include institutes of the *Centre National de la Recherche Scientifique* (CNRS), especially the four laboratories of its *Institut de Physique Nucléaire et de Physique des Particules* (known as IN2P3), and also laboratories of universities and other schools.

Most of these laboratories are contracted for research of the most fundamental type, often without immediate application, military or otherwise; but they also sometimes work on more direct military projects, which the defence clients often 'parcel' into various contracts, making the links quite difficult to establish.

In France, the CNRS is the major co-ordinator of all 'civilian' research, from humanities to physics and mathematics; it organises and partly funds hundreds of affiliated laboratories. Many well placed people (including in-house researchers) have no idea that the CNRS is engaged in work on behalf of the military. They are wrong – and have been for a long time. As early as 1939, the Caisse Nationale de Recherche Scientifique (which would become the CNRS later the same year) filed patents for the use of nuclear fission, including some specifically for atomic bombs.

In 1945 the CNRS, along with the new-born Commissariat à l'Energie Atomique (CEA) 'was part of the best established nuclear research institutions in the world'.[3] Throughout its history it has been heavily involved in fundamental nuclear and particle physics – creating in 1971 an affiliate institute, IN2P3, to co-ordinate work in this area. Today nearly 25% of the CNRS budget is devoted to nuclear-related research.

> A few years ago a group of CNRS researchers became suspicious about the origins of the contract for an atomic clock which they were developing. When they managed to get hold of the contract itself (which they had never seen before) they discovered they were working for the DGA on a clock which would become part of a missile guidance system.

Among the other laboratories referred to above are two which are specially interesting: the Collège de France (which has substantial facilities for nuclear physics research), and the Ecole Polytechnique which (aside from being militarily oriented and the most prestigious of the Grandes Ecoles) provides several hundred researchers to the Direction des Recherches, Etudes et Techniques (DRET) of the Ministry of Defence's Délégation Générale pour l'Armement, on its research projects. Indeed, 15% of DRET's contracts go to the CNRS or university laboratories.[4]

These research institutes also work under direct contract from the Direction des Applications Militaires (DAM) of the Commissariat à l'Energie Atomique (CEA), or from the DRET. But the DAM also conducts fundamental research on nuclear matters by itself through its six 'Centres d'Etudes', concurrently to manufacturing warheads. The DAM also benefits from research and documentation done by the *Institut de Recherche Fondamentale* (IRF) attached to the 'civilian' half of the CEA, and its four research centres.

All these laboratories undertaking fundamental research also have some input into the formulation of military research policy. They may propose general projects on their own initiative, and have some representatives sitting on committees advising the military and nuclear R & D co-ordinators. The IRF participates even more directly, through the co-ordination of the CEA General Administrator. Although it is an important part of the CEA, and its work is very important to the progress of French nuclear science, the IRF remains marginal to DAM policy initiation.

As might be expected, the information flow in these stages is strictly one way: DAM and DRET collect useful information from the Institut de Recherche Fondamentale and the independent laboratories but share none of their own.

CYCLE 3 – FUNDAMENTAL AND EXPLORATORY MILITARY RESEARCH AND PLANNING

A considerable amount of fundamental research is also conducted by the armaments community. These organisations and institutes also explore weapons ideas and initiate developments, as well as planning for their further progression in the development chain. The most important and influential of these organisations is part of the Ministry of Defence: the Délégation Générale pour l'Armement (DGA). Other organisations also contribute – the armaments corporations, and some public research institutions such as the Office National d'Etudes et de Recherches Aerospatiales (ONERA), and the national space programme the Centre National d'Etudes Spatiales (CNES) although this is not principally devoted to military activity. But the centre of everything is the DGA, which merits a detailed description.

The Délégation Générale pour l'Armement (DGA)

The DGA is responsible for the preparation of French armaments policy (but is not exclusively responsible for nuclear arms policy), as well as for overseeing the entire process of conventional arms and delivery systems procurement. For nuclear weapons, the DGA is not responsible directly for warheads and explosive charges – these fall to the CEA.

The DGA was created by De Gaulle in 1961 to rationalise and centralise the development, production, purchase and export of weapons; as such, it is one of the largest and most important state industrial enterprises in the country:

(1) it oversees the employment of 310 000 French men and women who work in the production of arms (of whom 75 000 are employed directly by DGA);
(2) it commands virtually the entire materials and equipment sector of the annual defence budget – about 50% of the global sum;
(3) the orders and contracts passed by the DGA account for 62% of the business of the French electronics industry, 70% of the aeronautics industry, 50% of the nuclear industry, 30% of the national R & D funding, and more than 50% of the budget of the CEA;[5]
(4) it trains its own specialists and élite staff by managing its own two 'Grandes Ecoles' – and has also specialised technician schools for lower rank specialists.

The scale and efficiency of the vertically integrated system is apparent in the rapidity with which France has evolved from a nation defeated first in World War II and then in two colonial wars, to become a formidable nuclear power and the largest per capita arms exporter in the world.

The DGA has a central role in the weapons procurement process. Its head – the Délégué – is among the highest ranking defence officials – on a par with the Chiefs of Staff of the armed forces. He reports directly to the Minister (in fact his office is closer to the Minister's office than that of the Chef d'Etat-Major des Armées!) and is represented at the most important committees related to defence development, including the Conseil de Recherches et Etudes de Defense (CRED), the administrative board of the CEA, the boards of the major research institutes, and the administrations of all the major armament industries.

In order to carry out its role as its own buyer, producer and merchant of arms, the DGA is divided into a series of departments and co-ordinating bodies, as shown in the organisational chart. Its role in weapons production will be examined separately in the next cycle concerning applied R & D and industrial development. Not all departments are central to our purpose, and only the most important are described in detail here: the co-ordinating bodies (Chargés de Missions 'Atome' and 'Recherche'); the key agency for pre-development studies (the Direction des Recherches, Etudes et Techniques); and the aeronautics and electronics research establishments (ONERA and SCEI).

The Co-ordinating Bodies

Within an enormous organisation such as the DGA, co-ordination with outside research and with the CEA (which develops nuclear explosives) is essential. The *Chargé de Mission 'Recherche'* is a leading scientist who advises the Délégué on research questions and acts as a link with the Minister of Research and Industry. The *Chargé de Mission 'Atome'* is a high ranking Ingénieur Général de l'Armement (IGA) expert in nuclear matters – he provides overall co-ordination with the Direction des Applications Militaires of the CEA (rather than formulating policies) and 'aids the Délégué in expressing the technical side of things in terms a minister (of defence) can understand'.

The Direction des Recherches, Etudes et Techniques (DRET)

DRET is responsible for the co-ordination of all non-warhead military pre-development studies in France. Acting upon the decrees of the Minister of Defence and according to the guidelines and budget constraints produced by the Conseil de Recherches et Etudes de Défense (CRED), DRET is one of the most active and important links in the technical chain of modern arms production in France.

In addition to performing a number of its own studies, it directly administers several major research institutions, co-ordinates 'military' activities in many others, and acts as a clearing house for information on any scientific development, civil or otherwise, which may have military applications.

Military R & D accounts for nearly a quarter of all scientific activity in France. In 1983 the military research budget was 19.1 billion French francs, more than a third of the public funds devoted to research.[6] It is

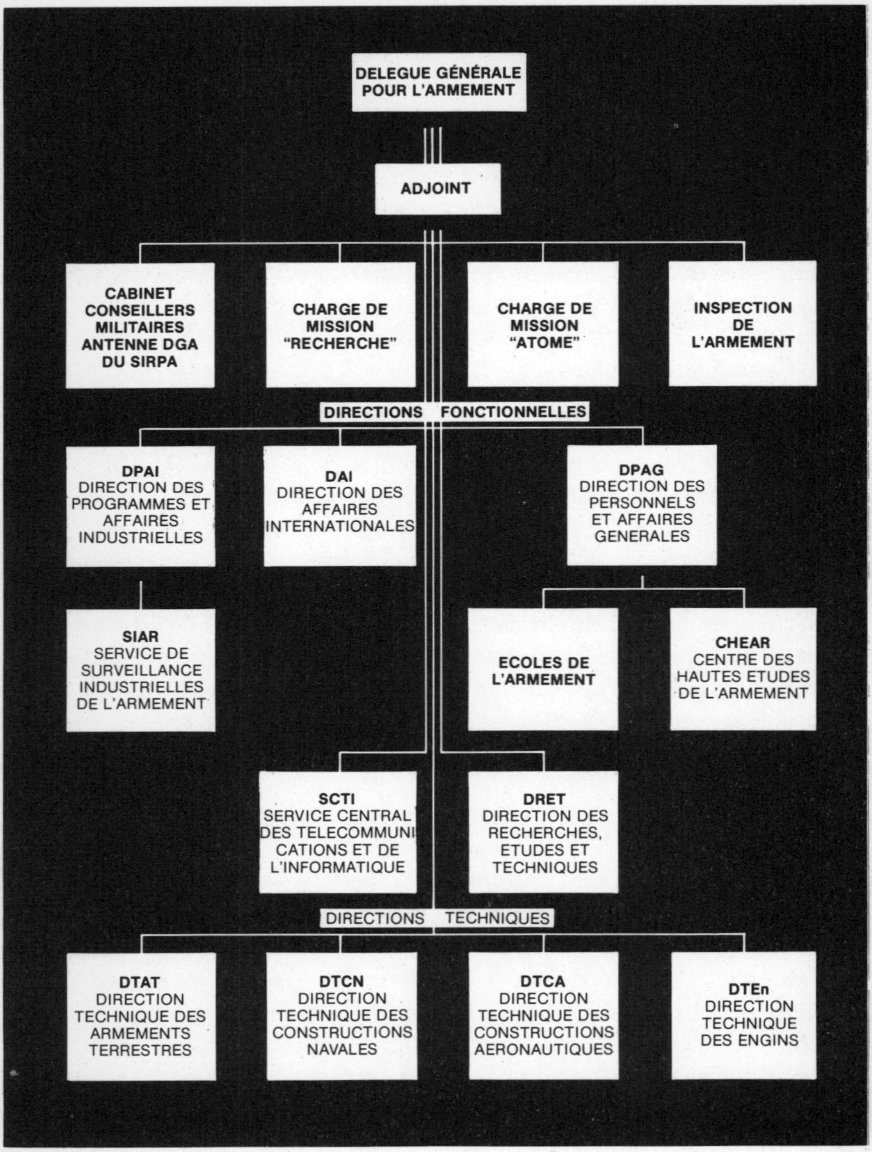

FIGURE 4.3 *Organisational chart of the Delegation Générale pour l'Armement*

difficult to know exactly how much of this sum is controlled by DRET. Almost 30% (5.9 billion francs) goes to pre-developmental work. This sum is shared between the CEA/DAM and DRET, with the bulk of it probably going to the latter. (In 1981 DRET passed contracts worth about 3.25 billion francs.)

DRET works in intimate co-operation with the armed forces, the CEA, the research community and the armaments industry. In particular, it works with GROUPE S (long term military/technical planning), CRED (planning of pre-developmental R & D), the Conseiller Scientifique au Ministre de Defense, the Mission 'Atome', and more directly with CEA/DAM (for the initiation of shared projects, development of hypotheses for weapons systems), the Etats-Majors, and the other DGA divisions. DRET's close contacts with industry stem from the fact that 73% of the research DRET sponsors is carried out in industrial laboratories.

DRET's role is basically pre-developmental. It has an entire department to help 'steer' inventions towards military applications. The Director of DRET has tended to be a specialist in the latest weapons 'fad' – for example, in particle beams.

> In commenting on the relative weight of DRET, one well informed source stated: 'DRET was far more responsible for the neutron bomb than was the Direction des Centres d'Etudes Nucléaires'; (DCEN is the military body involved with the CEA in the 'mise au point' of nuclear warheads).

The Office National d'Etudes et de Recherches Aérospatiales (ONERA)

ONERA is a scientific public establishment of an 'industrial and commercial character' operating under the direct authority of DRET. Working in close co-operation with the aerospace industry and with the DGA divisions related to aerospace, ONERA is the focal point for fundamental aeronautics research. In co-operation with the Centre National d'Etudes Spatiales (CNES) it contributes to R & D related to the military use of outer space.

In France, nearly all 'applied' aerospace R & D is carried out in the laboratories of the major industrial concerns (SNIAS, Matra, Dassault, and so on), while most research of a purely scientific or speculative kind is carried out in university laboratories and at specialised scientific establishments. ONERA is responsible for exploratory research (pre-'applied'); as such, it provides the link between pure scientific research and the aerospace industry. ONERA also provides one of the essential points of communication between the defence establishment and aerospace manufacturers. Integration with the (mostly nationalised) aerospace industry occurs at every level, from the administrative and scientific councils where the industrialists are represented, to the laboratories themselves where technicians from ONERA work side by side with technicians employed by private industry.

The Service Central de l'Electronique et de l'Informatique (SCEI)

This central electronics and computing service helps provide the access to modern technologies required by the defence establishment. Through specialised affiliated centres, it provides facilities for electronic testing and simulation of weapons systems (including very elaborate flight simulators), co-ordinates satellite communications programmes and advises the armed forces and government defence departments on the use of computers.

CYCLE 4 – APPLIED MILITARY R & D AND INDUSTRIAL DEVELOPMENT

The previous cycle of decision-making ends with the decision to pursue the final stages of development and to produce a weapon. Since the basic characteristics of the weapons have already been chosen, further decision-making concerns mainly technical co-operation. However, important changes in weapon design may still take place, and the policy governing the use of a weapon continues to evolve after the decision to develop has been taken.

Aspects of Nuclear Arms Procurement

The stages already described would provide a sufficient introduction to the R & D process for conventional arms. The procurement and R & D cycles for nuclear arms, however, are more complicated. 'Modernisation' of the nuclear forces involves improvements made by both the armaments production and warhead communities. Decisions to develop particular delivery techniques or particular charges do not depend critically upon one another until the decision to build a specific weapon. At that time technical co-ordination becomes necessary, but at earlier stages the armaments community is often given very little information about the nuclear charge that its missiles and planes will deliver.

The basic technology of nuclear explosions is now well understood and most current warhead research is 'fine-tuning' of old ideas.[7] Most newly-proposed nuclear weapons systems result from improvements in delivery capability (new basing modes, increased range, better penetration, multiple warheads, and so on). Certainly in many cases the improvements in delivery methods are only possible with improved warheads (especially those improved in miniaturisation and detonation accuracy). Few, however, are based on simultaneous conceptions of particular charge–delivery combinations.[8]

In general then, nuclear warhead R & D is separate from delivery system R & D. But they are both ultimately supervised by the Minister of Defence. One of the links is the Direction des Programmes et Affaires Industrielles, part of the Délégation Générale pour l'Armement.

The Direction des Programmes et Affaires Industrielles (DPAI)

The DPAI is responsible for the programming, co-ordination and implementation of the DGA's industrial policy. It is simultaneously a centre of planning, working with the Minister of Defence, the Etats-Majors, the Secretary General of Administration and others in developing overall procurement strategy and budgeting, and the main channel of communications between the DGA and the industrialists. It follows closely the industrial activities in the arms sector, and conducts price analyses. It can also initiate policy within the DGA.

Through its nearly independent affiliate, the Service de Surveillance Industrielle de l'Armement (SIAR), DPAI is in intimate and daily contact with industrialists at the level of individual factories.

SIAR is responsible for the daily control (mainly quality control) carried out within the arms factories themselves. Every major arms factory has a 'poste de contrôle' (there are 300 in all) composed of a

SIAR engineer and a small staff. They work out of offices on site provided by the factories. They act as the voice of the buyer to the factory, and the representative of the factory to the DGA. Thus, although the technical departments of the DGA (the Directions Techniques), work closely with industry on technical matters, political and industrial control is in the hands of the DPAI and SIAR, a situation not dissimilar to that in the Soviet Union.

Weapons Production and the Power of the Arms Industries

Weapons (excluding explosive charges of a nuclear nature) are manufactured in two ways: either by establishments and factories belonging to and run by the Directions Techniques of the Délégation Générale pour l'Armement; or by outside industries – mostly nationalised – which are contracted to the Directions Techniques and technically controlled and supervised by them – at least in theory. The Directions Techniques are closely involved with the arms planning process, advising the services and the Minister on the implications of technical developments and participating actively in applied R & D.

There are four Directions Techniques, each responsible for assuring the production of the arms in its domain:

(1) the *Direction Technique des Armements Terrestres* (DTAT), responsible for R & D and production of weapons for the Army. It is only involved in the nuclear arms development process through the production of the mechanical parts of ground launchers.
(2) the *Direction Technique des Constructions Navales* (DTCN) has its own ship building establishments and builds nuclear submarines at the Cherbourg arsenal; DTCN works in close contact with the CEA.
(3) the *Direction Technique des Engins* (DTEn) is the body charged with the development and construction of missile systems ranging from ballistic to tactical missiles. It works in tight collaboration with the firms building the missiles (SNIAS, Matra, GRIP, Thomson). It does not have its own production facilities but has important laboratories and testing capacities; and,
(4) the *Direction Technique des Constructions Aéronautiques* (DTCA) is responsible for the research, development and construction of military aircraft and associated systems. It does not have production facilities on its own – only testing facilities and research

assessment centres. Of all the Directions Techniques, DTCA might be said to be the least influential, largely due to the power of the large aviation companies.

The firms involved in manufacturing missiles and military aircraft are the most autonomous and the most successful of the French arms industries. Most of them are state industries, either nationalised (like SNECMA and SNIAS) or partly publicly owned (such as Dassault and Matra). These companies undertake most of the R & D in aeronautics and missile technology. There is also an intensive two-way movement of IGAs and military staff between the DGA and these large firms. They have considerable influence even in the earliest stages of weapons design; there is a constant interplay between the technical and industrial development of weapons by firms which, in principle, are responsible only for applied R & D and production, and the perceived needs of the military.

CYCLES 5 AND 6 – WARHEAD RESEARCH AND DEVELOPMENT

The French Atomic Energy Commission (CEA) is at the centre of all nuclear activity in France. It is responsible, either directly or through its subsidiaries, for the bulk of French nuclear R & D (including all of the nuclear military R & D), the production and handling of nuclear fuel at every stage of the fuel cycle, the development of French reactor technologies, the design, production, testing, and maintenance of all nuclear warheads, and the development of 'future weapons' such as particle beam guns and lasers.

The CEA is a mammoth organisation, having an annual operating budget of 15 billion francs and employing 36 000 people (including its subsidiaries) in 1983.[9] Although it is officially a public body under the authority of the Prime Minister and under the auspices of the Ministry of Research and Industry, the CEA has been described as a state within the state, and acts with considerable independence.

The CEA was created in 1945 by Charles de Gaulle. Later the Bureau d'Etudes Générales was set up within it to research and produce the first French atomic bomb. The innocent name of this bureau kept the activities of its members secret, even from highly-placed CEA executives. The 1950s saw the secret and rapid development of the bomb by the military half of the CEA, while the civilian half was working on reactor design and plutonium production. Then, in the mid-60s, as France was preparing to launch the most ambitious commercial nuclear power programme in the world, the CEA lost influence. A fight between the CEA's own reactor design, UGNN (comparable to the UK's Magnox) and the Westinghouse-designed pressurised water reactor (PWR), championed by the French electric utility EDF, was won by EDF. The French government's decision to adopt the PWR not only called the CEA's general role into question, but also demoralised the entire organisation.

In 1970, the CEA found itself criticised from all sides. Upon the advice of a special commission, the government ordered a complete reorganisation. Previously under the direct authority of the Prime Minister, the CEA was put under the Ministry of Research and Industry. Most of its nuclear fuel cycle activities were transferred to commercial firms, such as COGEMA.

Today, the CEA is making a recovery. The Socialist government has called for a national effort to increase the level of French R & D, and has committed itself to the modernisation of the nuclear forces. Both of these efforts involve a renewed role for the CEA. In 1982, the presidency of the CEA administrative committee (le Comité de l'Energie Atomique) passed back from the Minister of Research and Industry to the Prime Minister himself.

Accountability

M. Francis Perrin was High Commissioner of the CEA from 1951 to 1970. Now in retirement, he recalls that scant controls on the CEA's spending during that period made the Commissariat something of a scandal for the Finance Ministry. 'We were given a statute of autonomy analagous to that of Renault. The difference was that Renault was there to make money for the State, – and the CEA was there to spend it'.[10]

In 1982, the CEA was again re-organised and simplified. At the head of the administration remains the Administrateur Général, his adjoint and his immediate staff. The Administrateur is unquestionably one of the most important figures in the French nuclear system.

The 1982 reorganisation divides the CEA into four 'institutes' or divisions; the first three are far less central than the fourth to the nuclear weapons development process. The Institut de Recherche et Dévelopement Industriel is in charge of the co-ordination of industrial activity related to the fuel cycle and the construction of power plants. The Institut de Recherche Fondamentale is responsible for overseeing the 'civilian' research carried out by the CEA or in association with it. The third institute, the Institut de Protection et Sûreté Nucléaire is concerned with the physical safety and security of nuclear plants and with long term waste disposal. However, it is the Direction des Applications Militaires which is central for nuclear weapons decision-making.

The Direction des Applications Militaires (DAM)

This Direction is responsible for all of the CEA's military functions. It is a wholly independent department within the CEA whose activities account for almost half of the CEA's annual budget (44% in 1985[11]). The administrative structure of DAM, its internal budgeting, and the details of its activities are strictly secret. DAM employs between 8000 and 10 000 people. The engineers and scientists who run it are nearly all IGA's.

DAM carries out both fundamental research and industrial activities at six 'Centres d'Etudes' throughout France (Bruyères-le-Chatel, Limeil, Aquitaine, Ripault, Valduc and Vaujours). Information about the activities of these establishments is hard to come by. Implosion and detonation studies done for nuclear bombs are thought to be carried out at Ripault, which would thus be one of the key centres of neutron bomb R & D, and the secret military FBR, Rachel, is probably run by the DAM at Valduc.[12]

DAM carries out nearly all of the directly military research, both fundamental and applied, related to the production of nuclear warheads. DAM also manufactures the warheads, tests them in the South Pacific (at Mururoa), delivers them to the military, and helps maintain them on site at the military bases. DAM is not directly responsible for the production of the weapons grade nuclear material it uses. The plutonium, tritium, lithium and highly enriched uranium are all produced by Cogema. DAM simply places orders for the material it needs. The co-ordination of Cogema's activities with the needs of the DAM is a function of the Administrateur Général, and thus takes place at the highest level.

Recently there has been evidence that DAM may be reducing its activities. The basic arsenal for the Force de Frappe is in place, the R & D for tactical charges is probably complete, and DAM's latest major effort – the miniaturisation and detonation studies related to the neutron bomb – seems to have borne its fruit. While DAM still has a number of large tasks ahead of it (providing nuclear warheads for the new tactical missile Hadès, the medium-range air-launched missile (ASMP), the future ballistic submarine-launched missile M4 and so on) it seems to be trimming down its staff.

DAM clearly works closely with the Ministry of Defence and the Délégation Générale pour l'Armement. (The Ministry, in fact provides 99% of DAM's budget). Co-ordination between the two is provided by the Comité Mixte Armée-CEA, along with the Mission 'Atome' at the DGA, within which DAM works most closely with DTen and the main missile producer, SNIAS. Co-ordination at this level however is executive and fairly mundane. Co-ordination on policy takes place at high levels – between the Administrateur Général of the CEA, the Director of DAM and the Ministry of Defence represented by the DGA.

> As *Le Canard Enchaîné* has written, 'CEA scientists push hard for the bomb'. Clear examples of this exist in public statements by CEA officials in support of the neutron bomb, and their loud protests when the Socialists suspended (for three days) the bomb tests at Mururoa.

Nuclear warheads, of course, depend on the nuclear fuel cycle for their fissile materials, and the civil nuclear industry contributes to weapons developments indirectly, since it pays for much of the infrastructure behind the nuclear fuel cycle. Many argue that military programmes are now benefiting from the energy programme, just as the energy programme once depended on technical advances paid for by the military. This is specially obvious in the case of the FBRs.[13]

CYCLE 7 – THE ULTIMATE DECISIONS

The decision-making cycles described so far are responsible for the development of military and technical options. In the final procurement step the President and his advisers in the Conseil de Défense weigh the proposals of the defence community against the political and economic pressures constraining them. The decisions then taken open grounds for new avenues and new products to be explored by the technico-military complex.

Before turning to a brief analysis of the forces shaping these 'non-defence' constraints one remark is necessary; it should not be assumed that the proposals of the defence 'community' are ever presented with one voice. There are numerous competing sources of pressure: the engineers of the DGA, the Etat-Major Particulier (personal military advisers to the President whose advice weighs heavily, even though they are not themselves involved in the lobbies within the defence community), and the armed forces (themselves divided by inter-service rivalries). The nuclear community also exerts an independent influence directly on the decision-makers. Thus, among the political constraints affecting the politicians is the need to consider the rivalries among the bomb producers and 'users' themselves. Since these men, particularly in the army, are all capable of affecting public opinion, these considerations may be very important.

Nevertheless, the bodies and interest groups concerned with nuclear weapons development share many assumptions in common, and they have been little disturbed by opposition from the political parties or the public. The idea that French nuclear weapons make France independent of both superpowers gives the concept of deterrence considerable ideological weight. The parties of the left, originally bitterly opposed to this central tenet of Gaullist policy, now rely on it extensively. The trade unions, which are real political forces, hardly debate defence policy. The public attitude to nuclear arms issues is predominantly one of passive acceptance or disinterest. As a result there is a wide-spread consensus on the need for nuclear weapons. This facilitates the secrecy of the decision-making process, and contributes to the concentration of decision-making on nuclear weapons within few hands.

The Conseil de Défense

The Conseil de Défense, also known as the Comité de Défense, is the highest government policy-making body dealing with national defence.

The Conseil is presided over by the President of the Republic, and conducts its business in absolute secret. Its official members are: the Prime Minister; the Ministers of Defence, Foreign Relations, Economy and Finances, Research and Industry and the Interior; the four Chefs d'Etats-Majors of the armed forces; the Secrétaire Général de la Défense Nationale (SGDN), who acts as secretary to the Conseil; the

Secrétaire Général de l'Elysée; and the Chef de l'Etat-Major Particulier. The Conseil has the right to invite *ad hoc* members to join any particular meeting, and often does. In fact, there are currently semi-permanent members who are very official – and utterly anonymous.

While the President has the authority to make the final choices, it is certain that the Conseil is the most important working group involved.

> The way the Conseil works is exemplified by its meeting of 30 October 1982. The fact that the meeting had taken place was a secret until two weeks later when Hernu (Minister of Defence) and Mitterrand made the announcement that shocked many of their supporters: that France was to modernise her nuclear forces at least as fast as the Right had planned prior to May, 1981. Specifically, the green light had been given for the seventh ballistic missile submarine, the new tactical weapon Hadès, and the development of the SX mobile missile. Later, faced with the stunned queries of some of his own party members, Mitterrand would explain that he was acting on the basis of 'new information' about the international situation made known to him by French intelligence during the course of Conseil meetings over the summer of 1981.

Since the Ministers of Finances, Research and Industry and Foreign Relations are among the most important participants in the Conseil this section includes a brief discussion of their role.

The *Ministry of Finances* is probably the single most powerful of all the French ministries. It is in charge of taxation, customs, the treasury, the budget, financial review of the workings of government, and the enormously important governmental control over the industrial sector, through its omnipresent controllers and inspectors. The role of this ministry is to exercise financial control over government spending. Its actions cannot help having a political effect, but its primary orientation is toward the apolitical, financial issues. In the field of nuclear weapons development, in fact, the only real opposition and debate has come from this ministry. During the period when France was in the process of committing itself to the world's most ambitious nuclear energy programme, the only voice of doubt within the circle of decision-makers was that of the Ministry of Finances. Even today, if the nuclear programme is to be pared down, or the expansion of the defence budget to be restrained, it is likely to be as a result of the pressues of economy.

In October 1982, Mitterrand cut several billion francs out of the defence budget (including about 10% of the 1982 allocation for nuclear weapons). His opponents accused him of bad faith and of having a weak defence policy. But everything about Mitterrand has shown that he has no qualms about spending money (and making money) on weapons. According to military commentators, the force behind the 1982 cuts was 'les Finances'.

The *Ministry of Research and Industry* is an important ministry charged with overseeing the nation's industrial and general scientific efforts – two realms currently being given top priority. Although it is a large and active ministry, its influence in nuclear weapons decision-making is limited. It has some role in the industrial control of the armaments industry. It has some budgetary authority and several important responsibilities in the field of research. It is nominally responsible for the nation's energy policy and the supervision of the energy industry, including the CEA and Electricité de France (EDF).

The infuence of the *Ministry of Foreign Relations* has grown since the departure of de Gaulle, who took all decisions on foreign policy. Also, France has been moving to a point in which European questions have become more important, and diplomacy more sensitive. Moreover, defence issues have begun to play a larger role in foreign policy. Within the Ministry the key department is that of Political Affairs, in which there are two sections of interest: the Office of Strategic Affairs and Disarmament, and the Office of Atomic and Outer-Space Affairs. The responsibilities of the first seem to be mainly involved with disarmament negotiations and the international situation.

The President of the Republic

The President presides over the Conseil des Ministres and the Conseil de Défense. He alone has the authority to unleash the French nuclear forces. While the President makes the final decisions, he makes them on the basis of options and advice presented to him by others. At the Elysée itself there are several offices which have a direct effect on what the President hears about nuclear weapons decisions.

Le Cabinet du Président is, of course, highly important – although at the President's level his most cherished advisers are far too important to be among his cabinet. The cabinet is a technical advisory group, which filters the information reaching the Elysée every day and reduces it to a manageable quantity.

The Etat-Major Particulier, is a group of high ranking military whose advice on policy is extremely influential. In May 1985, its head, General Saunier, considered one of the men who has the President's ear, was named Chef d'Etat-Major des Armées. These military men are above all familiar with the doctrine, the tactics, and the equipment of the Force de Frappe. The Chef is usually an officer of the Armée de l'Air who has commanded the Strategic Forces. The Etat-Major Particulier is responsible for the preparation and management of the daily engagement codes which the President would use to order a nuclear strike. One French defence scholar called the Etat-Major Particulier 'the center for applied strategic thinking'. Its members are part of the military intellectual elite and among those most intimately involved with the practical aspects of the nuclear arsenal. They are the military whom the President sees most often.

The Secrétaire Général, whose main function is the preparation of presidential decisions and orders, has some influence over what the President hears. He is most linked with the military matters through his Chargé du Renseignement who acts as the contact point at the Elysée for the various French intelligence agencies. The Secrétaire also attends meetings of the Conseil de Défense.

The President has many other advisers, and there is constant speculation about who holds most influence. The offices mentioned above are however the most important ones actually on the President's staff.

Secrétariat Général de la Défense Nationale (SGDN)

SGDN is attached to the office of the Prime Minister. It employs around 600 people, three-quarters of whom are on 'service détaché' from the Ministry of Defence. It has divisions and bureaus dealing with such fields as intelligence, nuclear affairs, military R & D and strategic policy planning, and its activities bring it into high level contact with the military, the President and the most important ministers. The Secrétaire Général is always a military officer of high rank, and his adjoint is usually a diplomat. Officially, the SGDN's job is to assist the PM in all of his military functions. As we have said elsewhere, the PM is a marginal actor in defence matters. It is no wonder that his assistant, the SGDN, is marginal as well.[14] The SGDN is not a source of any real policy, nor even of important executive decisions.

SGDN also works with two external bodies of some interest:

(1) The *Comité d'Action Scientifique de la Défense* is a committee composed of eight members named by a number of different ministers. It is meant to keep itself in contact with all of the important research organisms, and especially with the DGA. In fact, there is no evidence that this committee is very active.[15]

(2) The *Institut des Hautes Etudes de Défense Nationale* (IHEDN) is perhaps the most interesting body associated with the SGDN. It is an annual think-tank attended by a hundred or so people from industry, government, the military, the labour movement and the scientific community. Men and women who have risen fairly high within their profession are invited. The think-tank takes the form of a seminar taking place weekly throughout the academic year. The participants are encouraged to speak very candidly about their opinions on defence issues. The debates and the final report of the seminars are kept secret, although the recommendations are shared with the government.

This may be one of the only places where decision-makers are confronted face to face by people holding views radically different from their own, and where various factions of decision-makers are confronting each other in an open way. It is a fascinating, and perhaps very French phenomenon that a nation which has stifled its own public defence debate has set aside an academic setting where members of various kinds of elites can argue their positions.

IHEDN has often been the forum for important speeches describing the government's defence policy. Generally the Prime Minister or the Minister of Defence delivers the opening address, using the moment to outline the government's defence manifesto.

Intelligence

Information gathered by the intelligence community flows directly to the President as well as through the interpretation of the Ministry of Defence. In France the intelligence community is not nearly as influential in nuclear arms policy as in the United States and the Soviet Union.

Nevertheless, Mitterrand explained his volte-face on the arms issue as the result of intelligence information about Soviet activities which he learned only upon his arrival at the Elysée.

On the edges of the political stage are several bodies which act as minor counterweights or controls. Of these, the most important is the Parlement, and even its role is at best peripheral.

Parlement

One of the leading scholars of French constitutional law has written: 'The powers of the French Parlement are more reduced than those of any other parliament in the West – especially its legislative powers'.[16] This constitutional weakness is compounded by the lack of information and unquestioning conformity that characterizes the 'silent consensus' surrounding nuclear arms and nuclear energy issues in France. The Parlement has had neither the power nor the will to be a real decision-making force in these matters.

The Parlement is composed of two houses, the Assemblée Nationale (491 Deputés elected by universal direct suffrage every five years) and the Senat (283 Senateurs elected indirectly by local electoral colleges, serving nine years, one third of them being elected every three years). The indirect election of the Senat leads to a heavy over-representation of the provincial regions of the country: the Senat tends to be more conservative and more hawkish than the Assemblée.

In France the government retains broad legislative powers via ministerial decrees and orders. In fact, the legislative power of the Parlement is limited more to approving governmental projects, and defining the parameters within which the government may issue decrees, that it is an affirmative, policy-making legislative ability. To the extent that Parlement does legislate, its agenda is set entirely by the government. The government decides what Parlement will debate, and when. In Spring 1985 the debate on defence doctrine scheduled for April was simply cancelled by a decision from the Prime Minister.

The Assemblée is the stronger of the two houses, and could actually hinder the government if it had a mind to. The Senat has no power to block legislation which is supported by the government. The constitution is arranged so that the government may use the Senat to block legislation issuing from the Assemblée, but allows the Assemblée alone to enact laws with the government's consent. A relevant example of this occurred in late 1982 on the issue of manpower levels in the army.

Amid harsh criticism of the government from the military and the Right, the opposition-controlled Senat rejected the 1983 defence budget. It was the first time under the Fifth Republic that a defence budget had been rejected. Yet, despite the drama, the refusal by the Senat had no effect. The government simply enacted the budget after sending it back to the Assemblée for a second rubber-stamping.

The major means of independent action available to the Parlement is their right to question the government on its projects and actions. But since the Assemblée has little room in which to manoeuvre, even less than in England, the questions posed tend not to be too contentious. When it comes to the nuclear issues, no one seems to have any doubts to express.

> Parlement is neither a decision-maker nor a decision-stopper. It is, at most, a place to evoke debate. When the 1984–88 military programmation was voted after an afternoon of pantomimed debate, one leader noted blithely, 'Nous sommes tous d'accord'. In 1980, when the government revealed that the neutron bomb had been 'under study' for years, Parlement learned about it through the newspapers.

Both houses are divided into permanent and *ad hoc* commissions. The Commission de la Défense Nationale et des Forces Armées in the Assemblée, is responsible for preparing the debates and reports (Rapports d'Information) concerning defence issues. The president of this Commission has the power simply to reject any particular question for discussion. The Commission des Finances is the most powerful commission overall, and is responsible for the inspections and debates concerning budgetary matters – perhaps the only area in which real debate could eventually take place.

CONCLUSIONS

The powers of the President and of the various ministries in France are extensive, and their access to information is quite good. They are neither ignorant of the stages which take place earlier in the system nor without influence over them. The individuals in the most important positions all over the chart owe their places to government appointments. However, this does not generally make them *political* appointees. They are officials, and often survive changes of political leadership. Further, representatives of the government are found on all important advisory and administrative bodies involved in the process. These individuals share a set of assumptions and patterns of interpretation which cannot be fully controlled by government policy because the policy, and even the men who compose it, are a product of the system.

As unique to France as the lack of debate to which it contributes, is the existence of a pervasive, highly disciplined, bureaucratically oriented power élite – the mandarin system of the Grandes Ecoles and Grands Corps d'Etat. Except within certain pockets, the men in positions of real power throughout the system belong to some part of this caste of élite officials. This system, and the social attitudes which produced it, encourages conformity and stifles debate – especially with regard to issues 'de l'Etat' – of which defence policy is the very first.

NOTES

1. The fact that the DGA has recently pressed for a quota of 300 graduates of Polytechnique a year instead of the present 250, shows how much it values the services of Les X.
2. M. Debré, who was one of the authors of the *Constitution of the Fifth Republic*, was one of De Gaulle's Prime Ministers and became a Minister of Defence later on. Although a very forceful minister he was always under the orders of De Gaulle.
3. C. Sweet, *A Study of Nuclear Power in France* (Polytechnic of the South Bank, Energy Paper no. 2. London, Mar. 1981).
4. As reported in 'La Science au Service des Armes', *Ca M'Interesse* (Apr. 1985) p. 59, one out of every four of DRET's research contracts is awarded to the Compiègne University – a recently created 'high-tech' university of technology, the first of its kind in France; the government intends to create other similar ones.
5. *Défense – Revue de l'Union des Associations d'Auditeurs de l'Institut des Hautes Etudes de Défense Nationale*, no. 34 (Oct.–Nov. 1984) p. 20.
6. Ibid., p. 55.
7. Topics for current research include charge versatility (a free-fall bomb which bombadeers could set to different kilotonnages at the touch of a dial), charge miniaturisation, and 'special effects' (enhanced radiation). Only the last of these is a truly new weapon.
8. The neutron bomb is an exception to this generalisation. The neutron warhead (at least in its tactical uses) is only effective given certain precise delivery capabilities.

9. *CEA Rapport Annuel 1983.*
10. *Financial Times* (20 Feb. 1985).
11. Ibid.,
12. *CEA Rapport Annuel 1962*, p. 140.
13. For a discussion of this position, see M. Genestout and Y. Lenoir, 'Quelques Vérités (pas Toujours Bonnes à Dire) sur les Surgénérateurs', *Science et Vie*, no. 781 (Oct. 1982).
14. During the first decade of the Fifth Republic, the power of the SGDN was certainly more important than it is now; in 1969, when he was Minister of Defence, M. Debré forced the resignation of the Secretary and this changed the situation.
15. The DRET has published a very thorough description of the administration of military R & D in France: the Comité d'Action Scientifique de la Défense received no mention there.
16. M. Duverger, *Institutions Politiques et Droit Constitutionnel*, Tome 2 (Le Système Politique Français), Presses Universitaires de France, 15th edn (1980) p. 271.

5 China

The Chinese began their nuclear research in collaboration with the Soviet Union in the mid 1950s. However, by 1958 at the latest the Chinese had decided to develop their own nuclear weapons, since there were already problems in Sino-Soviet relations which led to a complete split by 1961. Considerable evidence exists to indicate that Soviet reluctance to transfer military technology, especially for nuclear weapons, was a cause of the Sino-Soviet split.[1] On 16 October 1964 the Chinese detonated their first atomic device and exploded a thermonuclear bomb on 27 December 1968.

Since that first atomic test, China has developed an increasingly varied arsenal of nuclear weapons. Although it does not approach the numbers of weapons held by the USA and USSR, China has already deployed at least a hundred nuclear missiles and is estimated to have over a hundred nuclear bombs for air delivery. This arsenal is being extended with the development of intercontinental and submarine missiles and tactical nuclear weapons. China is now in a position to threaten the European regions of the Soviet Union and parts of the Western United States, not to mention its nearer neighbours. China's ability to threaten the Western Soviet Union must inevitably affect the European nuclear chessboard. At the same time, any Soviet offer to move missiles from Europe alarms the Chinese if it means moving the missiles further East and nearer China. China cannot be left out of disarmament considerations.

The Chinese explain their development of nuclear weapons in terms of their deterrent value. At present the Chinese think the most likely attack on China would come from the Soviet Union. Most Chinese nuclear missiles are therefore in the North-West where they can threaten the Soviet Union. The Chinese must also take account of a possible nuclear threat from the United States (for example if the Chinese looked like invading Taiwan); they may have some missiles aimed at the USA. There may also be Chinese nuclear weapons designated for use or threatened use against neighbouring countries such as Vietnam (an ally of the Soviet Union) and perhaps Taiwan.

China's international strategy has changed dramatically over the last decade. In the 1960s, China saw both the Eastern bloc and the developed West as enemies. The Chinese press attacked the Soviet Union and Eastern Europe for 'revisionism', for backsliding on the revolution. At the same time America represented the evil of 'imperialism': American intervention in Vietnam seemed a potential threat to China's territory just as its participation in the Korean war had in the early 1950s.

1972 saw the beginning of a detente between China and the United States and, at around the same time, other Western countries. Increased commercial and diplomatic activity, academic and cultural exchanges, and tourism contributed to a more relaxed attitude both in China and the West. The 'yellow peril' was forgotten.

In the 1980s China has started to reach a similar rapprochement with the Soviet Union. Exchanges of all kinds have started to build up; the era of border clashes is over.

In South-East Asia, China has now abandoned its aim of fomenting revolution. Friendly relations with the maximum number of countries is now China's aim and practice, a policy which accords with her increasingly high profile in international bodies since her admittance to the United Nations in 1971. There have even been moves to establish a better rapport with countries like Israel and South Korea: trade is a key motive.

China's current preoccupation is with its own economic development. Under the new 'open-door' policy, Western companies have been encouraged to set up operations with Chinese partners in China. Closing the door to the West would certainly impede China's technological progress.

Under China's current leadership, the development of the economy has priority over military modernisation. Military research and production facilities have been required to produce goods in short supply for civilian use.

The nuclear weapons programme, however, continues to receive extensive support. It has been a priority project since its beginning in the 1950s, despite Mao Zedong's mockery of the atomic bomb as a 'paper tiger' and his avowed defensive strategy of sole reliance on 'people's war'.

Mao's view that nuclear war is inevitable and that China could easily survive a nuclear attack has also been modified. In 1980 it was announced that nuclear war could be postponed or prevented, and a number of pronouncements were made about the damage that nuclear war would inflict – 'the destruction of a future nuclear war will inevitably bring disaster to the whole world. There will be no victor but only massive global defeat'.[2]

The upheavals of the Cultural Revolution in the late 1960s caused chaos to production and research throughout China but did not disrupt the nuclear programme. Nuclear weapons have been regarded as good value for money and a question of national prestige by all Chinese governments, no matter whether their politics were 'radical' (like the 'Gang of Four') or 'practical' (as at present). In a major statement in October 1984, Defence Minister Zhang Aiping (who supervised China's first atomic test in October 1964) reaffirmed the importance of nuclear weapons to China in view of the limited finances available for military spending.[3] Greater efforts should be made, he said, to strengthen China's 'strategic counter-offensive ability'.

In a report based on US intelligence sources, the Far Eastern Economic Review[4] suggested that China was in the process of constructing ten new missile sites so that it would have a total of fourteen by mid 1984. Previously, according to the report, China had only four operational missile sites. China continues to have a strong commitment to the expansion of its nuclear forces.

THE PARTY, THE STATE AND THE ARMED FORCES

In the past, in China, major decisions were always presented as decisions of the Party. Lately, since the importance of State (or government) institutions was re-affirmed under Deng Xiaoping, Party and State institutions have in theory become more distinct. In a recent formulation the Party is described as being concerned only with matters of policy and principle (and personnel) while State bodies have the task of practical matters like administration and production.[5]

In practice the distinction is not so clear. The decision to initiate or renew a nuclear arms programme, for example, would be taken initially by the six men who form the Standing Committee of the Politburo (of the Party's Central Committee). Their decisions would then be expressed as policies of the government through the State Council to the bodies under it, such as the Ministry of Defence, or through the new

Central Military Commission (which is independent of the State Council). Figure 5.1 which follows illustrates this. In addition to working through State bodies, the Party leadership has its own command and information system to shape decisions at lower levels and respond to pressures from below.

Leading positions in the government are invariably held by senior Party figures, so that the systems interlock. The present premier Zhao Ziyang, for example, is on the Standing Committee of the Politburo. In recent years there have been calls for top figures to hold only one post at a time. China's strongman, Deng Xiaoping, who sits on the Politburo Standing Committee, seemed to be setting an example in May 1983 when he resigned from his chairmanship of the Chinese People's Political Consultative Conference (a political forum for people not in the Party) and withdrew from the National People's Congress (China's parliament, which meets once a year or less often). It is very significant, however, that Deng retained his powerful position as head of the Party's Military Affairs Commission and in June 1983 he became Chairman of the new Central Military Commission (which was apparently established to take over the functions of the Military Affairs Commission). His retention of these key posts is an indication of the continued significance of the military in decision-making in China. The top leadership of the armed forces, the People's Liberation Army (PLA), and the Party are similarly intertwined. The PLA falls directly under the Ministry of Defence, which is itself responsible to the new Central Military Commission.

In fact, major defence decisions probably rest in the hands of fewer than fifty individuals, many of whom occupy two or more key positions. There have been reports of other top Party leaders being reluctant to stand down from their controlling positions. There may, however, be a clearer separation of authority as the old guard die out and as Deng's reforms continue to define distinct areas of authority and activity.

There is no political force equivalent to the KGB in China. China's present leaders have a bitter distaste for any agency which might purge the Party from the outside since the majority (including Deng Xiaoping) were smeared and disgraced during the Cultural Revolution (1966–76). However, it remains to be seen how the new Ministry of State Security, established in June 1983 and charged with the task of 'ensuring China's security and strengthening the struggle against espionage', will develop.

The reforms of the past five years have been extensive. In May 1982 a

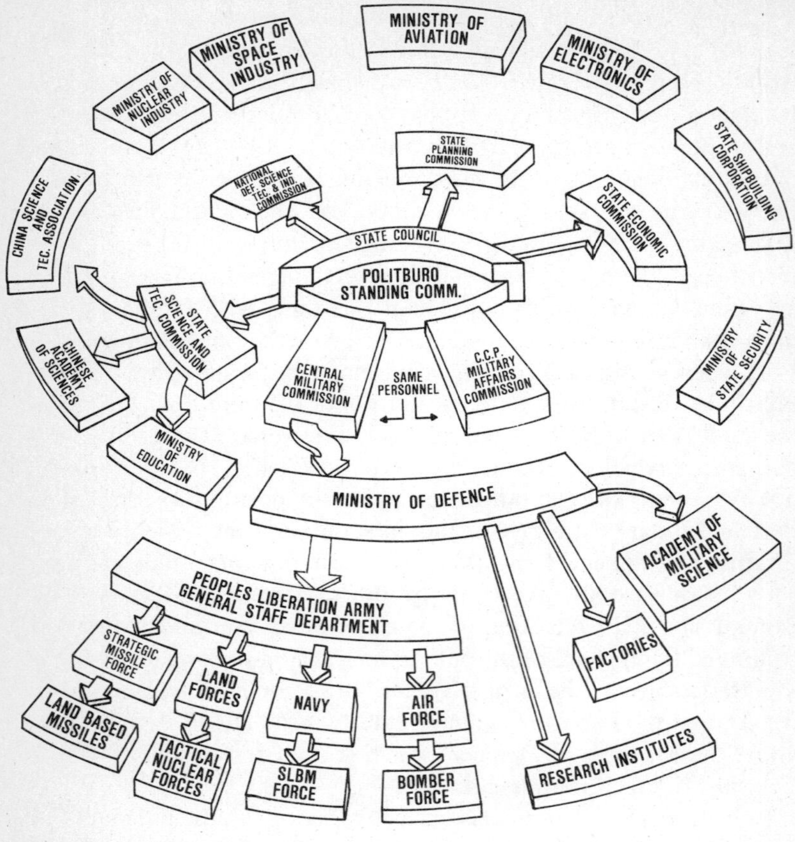

FIGURE 5.1 *The Party, the State, and the Armed Forces of China*

reorganisation reduced the number of ministries from fifty-two to forty-one. In September 1982 the Twelfth National Congress of the Chinese Communist Party adopted a new constitution, while in December a new State Constitution was adopted by the Fifth Session of the Fifth National People's Congress. Both entailed changes in the structure of power. Further reforms are expected.

The Sixth National People's Congress meeting in June 1983, named the new Head of State (a post reintroduced by the new constitution) as Li Xiannian. The restoration of the post has been interpreted as another move towards a separation of function between Party and State.

Centralisation, Bureaucracy and Discussion

The Communist government which came to power in 1949 inherited a tradition of bureaucratic attitudes in government left over from imperial times. Seventeen years later Mao Zedong's Cultural Revolution tried to create a more dynamic political climate, but finally hampered the economy and intellectual life since opinion makers and decision makers (officials, scientists, managers, journalists) became the targets of verbal and physical attack.

Today, Deng Xiaoping and his supporters see 'liberation of thought' as a key element in accelerating China's scientific and economic development. In the countryside, it is now up to the families to decide what to grow; the communes have been abolished; a significant proportion of peasants have become richer from the new policy. These successful reforms are now being repeated in other areas. They appear to be popular and are likely to outlive their architect, Deng Xiaoping.

Daily life for the average Chinese has altered dramatically since the reforms began in 1978 (following the death of Mao in 1976). The media, although Party controlled, is no longer confined to straight propaganda: there is variety and human interest; the main television news has extensive footage from foreign broadcasting agencies. There is less banner headline emphasis on the role of the army, which during the Cultural Revolution featured on the front pages almost daily. The military is however very much a part of Chinese life, with the armed forces engaged in public works and agriculture, and extensive militia training for civilians. There is a strong emphasis on civil defence, with foreigners regularly shown networks of underground shelters said to exist beneath factories and housing developments in many major cities.

The official press now frequently refers to the desirability of intellectual freedom. Moderately unorthodox opinions, provided they are expressed in the right way, can be voiced without fear of political retribution. In the last five years, travel abroad by decision-makers is broadening their perspectives. There are now press reports of debate and even the amendment of legislation by the National People's Congress, formerly a body whose role was merely to rubber-stamp decisions by Party organs.

Of course old habits persist. Although there is a climate where a limited debate is possible, there is little public awareness or discussion of nuclear issues. Key nuclear choices will continue to be made at the top.

THE COMMUNIST PARTY OF CHINA

At the very summit of the Communist Party are the six leaders who form the Standing Committee of the Politburo. They are (listed in Chinese protocol order): the Secretary General (the most senior official) of the Party (Hu Yaobang), the Chairman of the Central Military Commission (Deng Xiaoping), a Vice Chairman of the Central Military Commission (Ye Jianying), the President of the People's Republic of China (Li Xiannian), a former Vice Premier (Chen Yun) and the Premier (Zhao Ziyang). Hu Yaobang and Zhao Ziyang, the youngest members, are regarded as Deng Xiaoping's proteges; it can be assumed they are the most active in this Committee, most of whose members are in their 70s and 80s.[6]

The Politburo itself, which has twenty five to thirty members at any one time, acts as a sort of cabinet for the Party. All preparation work for the Politburo is provided by the Central Secretariat. Its Secretary-General is in a powerful position now that the Secretariat has been restored to its former importance in running the Party's day-to-day affairs under Deng Xiaoping.

The Politburo and the Standing Committee report to the Central Committee of the Party, but the Central Committee, although nominally the highest authority in the Party, rarely takes any initiatives and usually meets only once a year. The Central Committee is elected by the National Party Congress which is convened once every five years.

The Military Affairs Commission of the Central Committee was formerly a centre of power. It now seems to have been eclipsed by the new Central Military Commission, although it continues to be mentioned in the official press from time to time.

The top bodies in the Party dictate policy to the Government, through the Commissions and Ministries of the State Council; but the Party leadership have their own network of communication with all organisations of Chinese society through the Party committees which reach to shopfloor and village level everywhere. Policy decisions are announced and explained through this system and through the Party's newspapers and radio, which reach their massive public independently of the State administration.

The Communist Party has undergone considerable change in the last few years. Members have been required to renew their membership and the more radical or 'leftist' activitists are being weeded out. In their place the Party has accepted a large number of intellectuals, a group whose status has now been restored. The Party leadership has been instructing its local committees to keep out of state, legal and commercial matters, but this is evidently difficult for some local Party officials to accept.

THE CENTRAL MILITARY COMMISSION

The Central Military Commission was re-established in June 1983. (It had existed in the 1950s but was apparently abolished during the Cultural Revolution.) The Commission has the task of 'overseeing the Armed Forces', according to the Constitution.

It is thought to have taken over the role of the powerful Communist Party Military Affairs Commission. The previous leaders of this Military Affairs Commission seem to hold the same posts in the renewed Central Military Commission. The existence of the two parallel Commissions has puzzled the Chinese public: an article in the official newspaper for young people[7] explained that although the membership of the two bodies is identical, the duplication was necessary to keep Party and State bodies separate.

The Party Military Affairs Commission continues to exist but its functions are unclear. It would appear that the bodies formerly under it (the National Defence Industry Committee and the National Defence Science and Technology Commission) have now been replaced by the National Defence Science, Technology and Industry Commission, which reports directly to the State Council.

The Central Military Commission can be expected to formulate strategic military policy: it is probably the key body in deciding priorities within the defence budget. According to the Constitution, the Commission is not subordinate to the State Council but is responsible only to the National People's Congress.

The Central Military Commission is chaired by Deng Xiaoping, his most senior official position. It also includes the Deputy Secretary General of the Party, the Directors of the People's Liberation Army General Political Department and General Logistics Department, the Chief of General Staff and the Minister of Defence.[8]

THE STATE COUNCIL AND THE COMMISSIONS

China's parliament, the National People's Congress, is usually convened once a year, when it elects a Standing Committee and the State

Council. The Standing Committee acts for the Congress between its short sessions and approves legislation on its behalf.

The State Council is a cabinet of the ministers of all the ministries and equivalent bodies. The Premier, currently Zhao Ziyang, is head of the State Council.

In January 1983, the State Council set up a *Science and Technology Leading Group* to formulate long-term plans, evaluate policy decisions and exercise overall co-ordination of science and technology. The group is under the Premier, Zhao Ziyang. His two deputies are the Minister in charge of the State Science and Technology Commission and the Minister for the State Planning Commission. The group brings together interested parties from the relevant bodies.

Under the State Council there are a number of commissions which supervise and co-ordinate the activities of the different ministries concerned with the stages of nuclear weapons design, manufacture and deployment.

The State Science and Technology Commission. This has responsibility for the Ministry of Education and the Academy of Sciences. Nuclear matters are said to be the responsibility of the Commission's Fifth Bureau.

Ministry of Education. The institutes of higher learning which undertake theoretical and practical research on nuclear power and weapons fall under the Ministry of Education. It is thought that the National Defence Science Technology and Industry Commission advises the ministry on research priorities.

Chinese Academy of Sciences. The Academy, like the Ministry of Education, is under the overall supervision of the State Science and Technology Commission. According to a report in the Beijing Review,[9] the Academy has more than 110 research institutes under its central control. In addition there are several hundred institutes under local administration (for example the Nuclear Physics Institute, Shanghai). It is generally believed that the highest quality research is done in the centrally administered academies. Two of the essential institutions in nuclear weapons research are under central control and are in Beijing: the Institute of Atomic Energy, and the Institute of Mechanics, which is involved in rocket technology.

The Institute of International Strategic Studies, in Beijing, carries out studies on strategic doctrine, on which the Central Military

Commission and the Party secretariat can base their formulation of strategic policy.

In 1982 a new body was set up under the State Council for supervising civil nuclear power; it is not clear whether it is a separate body or a new sub-section of the State Science and Technology Commission.

The State Planning Commission allocates resources to the various ministries concerned with military production. Formerly it operated through the National Defence Industry Office but this has now been absorbed by the National Defence Science and Technology Commission. It is not certain how these bodies are related, but there must be considerable liaison between them.

The State Economic Commission is responsible for annual plans, while the State Planning Commission works on long-term planning. (In February 1980 a State Machine Building Commission was established to co-ordinate military and civilian work in the machine building industries; then in May 1982 it was merged with the State Economic Commission.)

The National Defence Science, Technology and Industry Commission has recently been returned to the control of the State Council after being subordinate to the Party Military Affairs Commission for several years. It was originally set up specifically to co-ordinate nuclear arms development and this must still be one of its main tasks. The importance of the Commission is that it is assumed to control the allocation of research and development funds between various defence projects. Although the finance and other support for research and development is provided by the ministries concerned, it is this commission which would determine the relative importance of research programmes and the size of their budgets. It also gives policy guidelines to the People's Liberation Army research institutes and liaises with the Ministry of Education and the Chinese Academy of Sciences on research priorities.

The Commission has also probably taken on the tasks of the former National Defence Industry Office, such as co-ordinating the flow of products between the factories and research institutes of the various ministries involved in military production and exercising some supervision of military factories under the Ministry of Defence.

Professional Bodies

There are a number of professional bodies under the China Scientific and Technical Association, which is, at least in part, subordinate to the State Science and Technology Commission. The Association includes the China Nuclear Society, the China Nuclear Fusion and Plasma Physics Society, the China Nuclear Physics Society, the China Nuclear Power Society and the Chinese Society of Isotopes.

These professional associations not only have the role of providing a means of personal contact between scientists in a particular field, they also provide a forum for debate on priorities and co-ordination of research.

Ministries Concerned with Armaments Production

There are currently six ministries concerned with the production of weapons, apart from the Ministry of Defence itself (which falls under the Central Military Commission). Until May 1982, they were known as the Second to Seventh Ministries of Machine Building. In May 1982, they were named according to their main sphere of interest. These ministries are involved in both civil and military programmes. Under the ministries are a number of research academies, but little information is available on them.

In the past, each ministry ran its own mines (where appropriate) and factories; in the case of the academics and research institutes, however, it is not known which institutions have been attached to a particular ministry at any particular time. (For example, at different times the First, Second, Third and Tenth Research Institutes have been identified as undertaking research for what is now the Ministry of Space Industry.) Later the National Defence Science and Technology Commission began to act as a broker between the ministries and institutes. It is not known whether this practice has continued after the reorganisation in May 1982. It may be that research institutes and academies have now been assigned to the ministries on a permanent basis.

The ministries are as follows:

(1) *Ministry of Nuclear Industry*. Composed of the Second Ministry of Machine Building and other bodies, the ministry has had control of all the stages in producing nuclear weapons. It has uranium mines at Maoshan and Zhushan in Jiangxi, Xiazhuang in Guangdong as well as at Tacheng in Xinjiang.

The main uranium separation plants are at Baotou in Inner Mongolia, Yumen in Gansu and Lanzhou in Gansu. The Lanzhou facility is said to produce most of the weapons grade uranium; it is supported by a related plant at nearby Helanshan. Warheads are reportedly manufactured at Huangyuan in Qinghai. Testing is carried out at Lop Nor in Western China.

> 'Since the first mushroom cloud arose over the Lop Nor area in Xinjiang region, the remote and uninhabited Gobi Desert, surrounded in a veil of mystery, began to attract the attention of the world. In mid-summer this year I visited this testing area, China's nuclear testing ground, which has been rocked by thunderous nuclear explosions.
>
> I departed by car from a place called Malan and headed for ruins of the ancient city of Loulan, I happened to travel in the same car with Zhang Zhishan, an old friend of mine whom I had not met for a long time. The former commander of the nuclear testing base, Zhang Zhishan told me that the Lop Nor nuclear testing ground, with a total area of more than 100 000 sq km is as large as Zhejiang Province. Nowadays in the testing zone there is a highway network with a total length of more than 2000 km as well as various facilities for nuclear tests to be carried out on the ground, on towers, in the air, by missiles, in horizontal underground tunnels or in vertical shafts. Command centres, telecommunications units, control centres, permanent monitoring stations and various facilities were built at all the test sites. A large airport and workshops for the manufacture of product-testing equipment are located somewhere far away. An odd scene finally presented itself before our eyes – dilapidated automobiles lay on rocks, armoured cars and aircraft had been turned into wreckage, dilapidated concrete buildings could be seen here and there and part of the surface looked like melted glaze. It looked as if a large scale modern war had been fought in the depths of this desert some time ago'.[10]

At a work conference of the ministry in January 1985, the Vice Premier who often speaks on nuclear matters, Li Peng, stressed that the nuclear industry, like other armaments producers, must help develop the civilian economy.

Under the ministry is the semi-commercial China Nuclear Energy Industrial Corporation, the body which deals with foreign nuclear co-operation.

(2) *Ministry of Aviation Industry*. Formed from the Third Ministry of Machine Building, the ministry is responsible for research and production of small missiles and aircraft, including the nuclear-capable B6 Hong aircraft (reportedly built at Xian and Ha'erbin), and the B5 (built at Shenyang).

(3) *Ministry of Electronics Industry*. Composed of the Fourth Ministry of Machine Building and other bodies, the ministry carries out research and production of electronic, telecommunications and navigational equipment.

(4) *Ministry of Ordnance Industry*. Formed from the Fifth Ministry of Machine Building, the ministry produces conventional weapons from rifles to tanks.

(5) *China State Shipbuilding Corporation*. Formed from the Sixth Ministry of Machine Building in May 1981, the corporation is concerned with both military and civil research and production in shipbuilding, which is presumed to include nuclear-powered and nuclear-missile submarines.

(6) *Ministry of Space Industry*. Formed from the Seventh Ministry of Machine Building, the ministry is responsible for R & D of ballistic missiles and satellite launches, and is also said to oversee the guided missile programme, in which case it may include the former Eighth Ministry of Machine Building (formed in September 1979). Major missile test sites (assumed to be under the control of the Space Ministry) are said to be at Shuangchengzi in Gansu and Wuzhai in Shanxi.

MINISTRY OF DEFENCE AND PEOPLE'S LIBERATION ARMY

The Ministry of Defence falls under the State Council and the Central Military Commission. Under the ministry is the General Staff Department of the People's Liberation Army (PLA) which links the various service branches; in the case of nuclear weapons, this involves the Strategic Missile Force, the Air Force, the Navy and the land forces.

The Strategic Missile Force, (formerly the Second Artillery) controls land-based missiles of four main types:

(a) The Dongfeng T-1 (or East Wind, called in the West the CSS-1) was the first Chinese strategic missile. It was developed from the Soviet SS-2 Sibling, and was first tested in 1966. Now considered unwieldy since it can only be fuelled immediately before launch, its range is estimated at about 1100 km, and it is believed to carry a single 15–20 Kt warhead. Between fifty and ninety are now deployed.
(b) The CSS-2, first deployed in 1971, has a range of 3000 to 5000 km (depending on the version) with a single 1–3 megaton warhead. Between twenty and sixty are now deployed.
(c) The CSS-3 was deployed in the late 1970s. It has a range of approximately 6500 km and carries a single 1–5 megaton warhead. Only four are believed to be in service; this missile was perhaps developed in case the more elaborate CSS-4 failed.
(d) The largest and most recently developed missile is the CSS-4, which is based on a satellite-launch vehicle (the Long March CSL-2), first tested in 1980. Its range of 11 000 km means it could reach Moscow and the Western USA. It is believed to carry a single 3–4 megaton warhead. It is thought that two and possibly as many as ten are already in service (deployed near Sundian and Tongdao, according to US intelligence sources).[11]

The Air Force probably controls the PLA air delivery nuclear force:

(a) The B6 Hong (similar to the Soviet Tupolev TU-16 Badger) entered service in 1968. Used in several nuclear explosion tests, it is believed to carry a 1 megaton bomb and have a range of 6400 km. About ninety are in service.
(b) The B5 (similar to the Ilyushin IL-28 Beagle) has a range of 2000 km and could carry a 1 megaton bomb. About 400 are in service.
(c) The F-9 Fantan (a modification of the MiG 19) could be used for tactical nuclear weapons.

The Navy controls China's submarine launched nuclear potential. On 12 October 1982 China launched a 1600 km-range missile (the CSS-NX-4) from a submerged submarine (probably a Golf-class) and can be expected to have SLBMs operational in the near future. China also has a nuclear-powered Han-class submarine.
Tactical Weapons are under the control of the Strategic Missile Army Units and the Air Force. Nuclear tests observed in 1977–8

indicated the development of low-yield tactical nuclear weapons. A recent Chinese military exercise also simulated using such a weapon. These would probably be delivered by FROG-type missiles, the Fantan fighter or the B-5 bomber.[12]

Military organisation in China is so compartmentalised that co-ordination between the services is believed to be almost non-existent. There is very little cross-fertilisation between units of separate services. Large scale combined manoeuvres, for example, are rare. The use of weapons by the individual services is quite separate with little co-ordination and as far as is known each service tests its own nuclear delivery system separately. In terms of manufacturing oversight, a weapons system required by a particular service would be debated by the Central Military Commission, which has the ultimate authority to prioritise which weapons are produced. The decision would then pass through the National Defence Science, Technology and Industry Commission, which has the authority to allocate resources for military programmes to the six ministries concerned with defence production.

Shadowy debates over military spending surfaced during the early 1960s, early 1970s and late 1970s. Although there was agreement on the importance of the nuclear programme, there were arguments between supporters of 'electronics' (high technology) versus 'steel' (heavy industry) in the national and military budgets (in the early 1970s). In the early 1960s and late 1970s there were those who advocated making heavy industry and military investment the 'key link' in the economy (to the detriment of light and other industries), while others believed that military modernisation should be treated as only one part of the whole economy and as a dependent part of it. The Four Modernisations (agriculture, industry, national defence, science and technology), popularised under Deng Xiaoping's leadership since 1978, gives priority to the development of the national economy as a whole. This has meant a curb on military expenditure.[13]

According to Chinese budget figures, defence spending, which was running at US $13 000 million per annum in 1979 was cut by a sizeable proportion to US $10 000 million for 1980–81, rising only to US $10 500 million for 1982–83. Although estimating and analysing the Chinese defence budget is notoriously difficult, there is agreement that an overall reduction in military spending did take place.[14]

The armed forces, currently estimated at up to four million men, are being streamlined and their manpower reduced. Party Secretary General Hu Yaobang told a New Zealand press conference in April

1985 that up to a million soldiers would return to civilian life over the next two years.[15]

There is resentment within the armed forces over these reforms and considerable 'leftist' displeasure at the shrinking importance of the PLA in Chinese politics and social life. Deng has removed from the PLA the political role it played during the Cultural Revolution as guardian of the revolution: its sole purpose now is to defend the country. At the same time, the new uniforms and spirit of professionalism introduced by Deng are held to have won him considerable support.

DISARMAMENT, PROLIFERATION AND NUCLEAR ENERGY

Since 1964 the Chinese government has repeatedly proclaimed that it will never be the first to use nuclear weapons. However, China has not signed the 1963 Partial Test Ban Treaty nor the 1968 Non-Proliferation Treaty (although it has indicated that it will act as if it had signed the 1968 Treaty). The Chinese have argued that the two treaties have been used by the superpowers to ensure that their dominant position in the world continues, at the expense of other nations' interests. China insists that 'Every country is entitled to participate on an equal footing in the discussions and negotiations for a solution of the disarmament problem'[16] and not just the nuclear powers. The Chinese have been supporters of nuclear-free zones and in 1974 signed the Treaty of Tlatelolco which prohibits nuclear weapons in Latin America. Wishing to appear as the champion of the Third World, China has adopted the (rather convenient) position of calling on the super-powers to make the first moves towards nuclear disarmament; China promises that, if the super-powers will first halt all development and halve their nuclear armouries, China will stop testing and production of nuclear weapons.

Press reports have suggested that the Chinese have given Pakistan help in developing a nuclear bomb. This has always been strongly denied by the Chinese. In an interview in January 1985, Vice Premier Li Peng said 'I would like to reiterate here that we have no intention, either at present or in the future, to help non-nuclear countries develop nuclear weapons'.[17]

China has no official peace movement similar to that in the Soviet Union and there is no unofficial activity either. (The 'spring' of 1978–79 when spontaneous posters appeared on Democracy Wall and unofficial groups emerged has since been suppressed). However, the official press has recently expressed some sympathy for anti-nuclear movements and a group of Chinese observers attended the Japan Congress against Atomic and Hydrogen Bombs in August 1983. Before the Japan Congress the following year, a peace rally was held in Beijing, in July 1984, organised by the Chinese Association for International Understanding. In the past the official Chinese view of anti-nuclear protest was that it weakened defences against the common enemy of China and the West, the Soviet Union. Chinese relations with the Soviet Union have recently improved.

As far as peaceful uses of nuclear energy are concerned, China is planning a number of nuclear power plants, the most well-known being that planned in collaboration with France and the UK (probably with GEC supplying the turbines, and Framatome the reactor) in Guangdong Province, near Hong Kong.

China plans further nuclear power plants, some in co-operation with foreign companies. A West German consortium (Nukem) has even discussed the possibility of storing foreign nuclear waste in the Gobi Desert. There is a growing exchange of business and scientific delegations with foreign countries. China has nuclear co-operation agreements or exchanges with Britain, France, Japan, Pakistan, West Germany, Brazil and the USA. The USA at present has legal obstacles to selling nuclear technology and equipment to China until China allows independent inspection of its nuclear facilities. China has recently joined the International Atomic Energy Agency but, like other nuclear powers, seems likely to refuse inspection of military installations.

NOTES

1. H. Ford, 'Modern Weapons and the Sino-Soviet Estrangement', *China Quarterly*, no. 18 (Apr.–June 1969) pp. 160–173.
2. Beijing Review, quoted in *China Now*, no. 111 (1985).
3. *Guardian* (5 Oct. 1984), *Beijing Review* (29 Apr. 1985).
4. *Far Eastern Economic Review* (9 June 1983).
5. BBC, *Summary of World Broadcasts, Far East* (18 July 1983).
6. W. Bartke, *Who's Who in the People's Republic of China* (Hamburg: Harvester, 1981).
7. *China Youth Daily* (23 July 1983).
8. W. Whitson (ed.), *Military and Political Power in China in the 1970s*, (Praeger, London: 1972); H. Jencks, *From Muskets to Missiles: Politics and Professionalism in the Chinese Army 1945–81* (Colorado: Westview Special

Studies, 1982).

9. *Beijing Review* (14 Feb. 1983).
10. Text of report by Guo Cheng 'A visit to Lop Nor nuclear testing ground', reported by *China News Agency* in Chinese (15 Oct. 1984).
11. *The Economist*, Foreign Report (1 Nov. 1984).
12. *Far Eastern Economic Review* (23 July 1982 and 16 Oct. 1982), and *The Economist*, Foreign Report (1 Nov. 1984).
13. P. Godwin, *The Chinese Defence Establishment* (Westview, 1983) p. 44f.
14. Ibid., p. 53.
15. *Guardian* (20 Apr. 1985).
16. BBC, *Summary of World Broadcasts*, Far East (5 May 1983).
17. Ibid. (23 Jan. 1985).

6 North Atlantic Treaty Organisation

THE STRUCTURE OF NATO

NATO is an alliance of sovereign nations. It was formed, and it is sustained today, as the basis for defending its member states against perceived threats to their security, independence and territorial integrity. Different political interpretations of NATO's actual role tend to depend upon the perspective of the interpreter. Member governments see it as a defensive alliance for securing peace. The Warsaw Treaty nations see NATO as a threat, part of a hostile encirclement of the socialist commonwealth, and as a bastion of American imperialism.

NATO is also a vast and diverse complex of political, military, administrative and executive bodies, with its headquarters in Brussels and with institutions and installations all over Europe and beyond. The organisation is far more than the sum of its individual members: it has an independent identity as a fully matured intergovernmental organisation, embracing hundreds of committees, employing thousands of personnel, with a never-static agenda of tasks to be fulfilled, plans to be designed, jobs to be carried out, goals to be assigned.

In order to understand how NATO really works, and how decisions are really made, it is necessary first to grasp the formal structure.

The North Atlantic Treaty

The Alliance is constituted round the North Atlantic Treaty, which was signed in Washington on 4 April 1949 by twelve governments: those of Belgium, Canada, Denmark, France, Iceland, Italy, Luxembourg, Netherlands, Norway, Portugal, the United Kingdom and the United States. Since then, four more countries have acceded to the Treaty: Greece (1952), Turkey (1952), Federal Republic of Germany (1954) and Spain (1982). Although there is a mechanism for withdrawing from the Treaty, no government has ever done so.

The central principle of NATO is found in Article 5 of the Treaty: 'The Parties (to the Treaty) agree that an armed attack against one or more of them in Europe or North America shall be considered an armed attack on them all . . .'. This is the alliance principle of collective security. NATO is an inter-governmental, not supranational, organisation, in which member states retain their full sovereignty and independence. NATO links these governments together, but it has no compulsory binding power over them. The defence policies of NATO nations are national responsibilities: the function of NATO is to provide an organisational framework in which national policies are co-ordinated for the collective good, and in which nations are able to consult, deliberate and reach common diplomatic positions on issues of general concern. For descriptive purposes NATO organisation can be divided into three areas:

Diplomatic/political organisation;
Military organisation;
Bureaucratic/administrative/executive organisation.

Diplomatic/Political Organisation

THE NORTH ATLANTIC COUNCIL

The highest forum for consultation and decision in NATO is the North Atlantic Council, a diplomatic assembly composed of ambassadors of NATO countries. It operates at two main levels. Firstly, each nation posts a Permanent Representative or ambassador to NATO headquarters in Brussels, Belgium, to represent the national position in Council, meeting weekly, and to conduct NATO diplomatic business on a full-time basis.

Secondly, twice a year the North Atlantic Council meets at senior Ministerial level, involving national Ministers of External/Foreign Affairs, and sometimes national Defence Ministers. These NATO Council Ministerial Meetings usually take place in a national capital, last around two days, and conclude with the issue of a public communiqué.[1]

THE DEFENCE PLANNING COMMITTEE

In terms of seniority the NATO Defence Planning Committee (DPC)

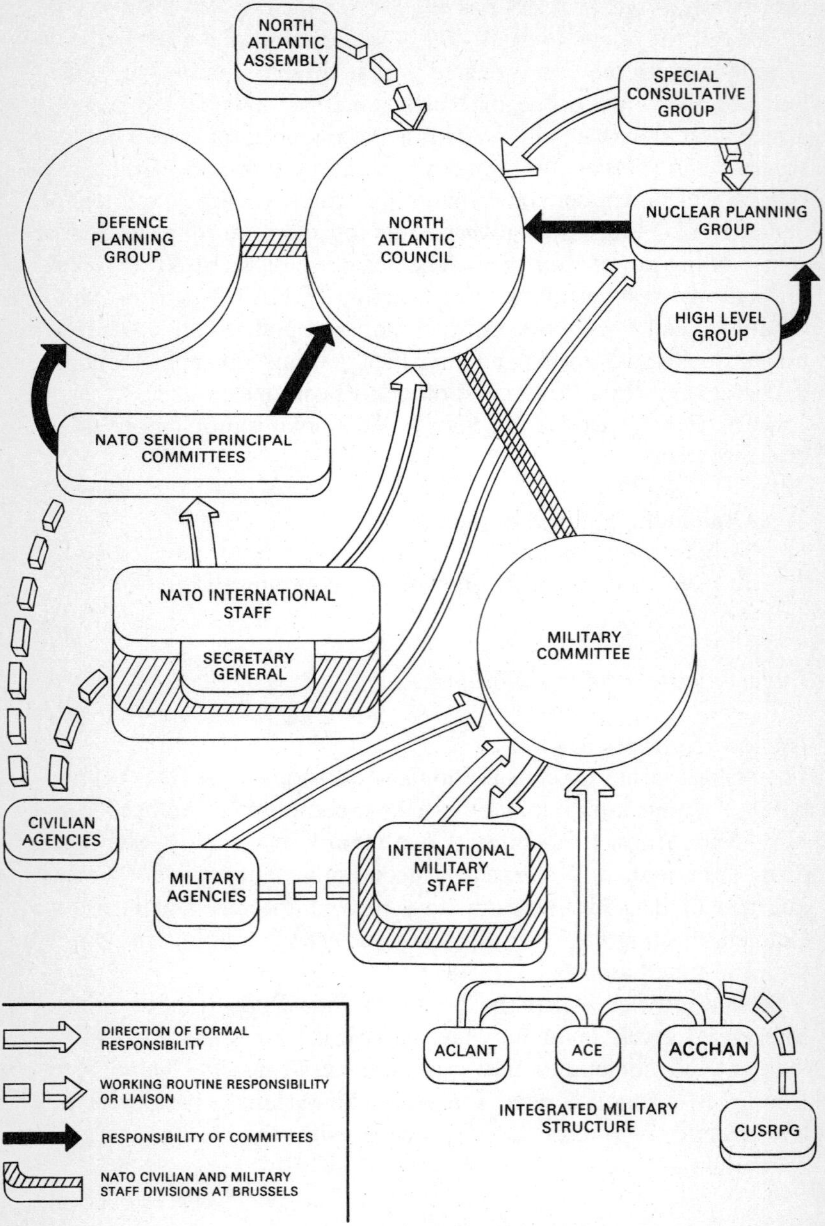

FIGURE 6.1 *The Structure of NATO*

has equal status and authority to the North Atlantic Council. It convenes twice-yearly at the level of national Ministers of Defence, and its decisions are supreme expressions of NATO policy on military force planning, procurement and nuclear affairs.

THE NUCLEAR PLANNING GROUP

The NATO Nuclear Planning Group (NPG) is the offspring of an earlier Ministerial committee, the Nuclear Defence Affairs Committee, which was set up in 1966 to deal specifically with the nuclear weapons policies of the Alliance but which, for all practical purposes, no longer exists.

The NPG meets twice-yearly at the level of National Ministers of Defence, to review Alliance nuclear policies, to frame decisions, and to consult on nuclear affairs. Its decisions are passed up to the DPC for endorsement. Its meetings last one or two days, and end with the issue of a public communiqué. The NPG also convenes at two other levels: members of the permanent Brussels delegations meet, about once a month, in the NPG Permanent Representatives Group. At a lower level NPG member nations meet in an NPG Staff Group under the auspices of the NATO Nuclear Planning Directorate, which is part of NATO's international staff.

Since its creation until March 1985 the group met thirty-seven times, and its major decisions have been approved by the North Atlantic Council and the Defence Planning Committee to become NATO's collective position.

What must be emphasised is that the term 'Nuclear Planning Group' is a misnomer. The NPG doesn't plan and it isn't just a group. It is a consultative forum, a place for joint deliberation. Consultations of one sort or another, at some level in the nuclear staff hierarchy, are going on all the time. A better title for the NPG is the 'NATO Nuclear Consultative System'.[2]

THE HIGH-LEVEL GROUP

In 1977 the NPG took the unusual step of setting up a new sub-committee of senior personnel from national defence ministries to prescribe detailed nuclear policy requirements for the future. This group has met regularly since that time and is called the High-Level Group (HLG). The structure, processes and recommendations of the HLG are very significant for NATO's nuclear affairs today, but the existence of this group is not acknowledged in the *NATO Handbook* or in the *NATO Facts and Figures* official volume. The HLG's activities are unpublicised and no communiques of its meetings have ever appeared. It is chaired by the US Assistant Secretary of Defence for International Security Affairs.

The High-Level Group has become NATO's nuclear weapon think-tank: a committee of senior defence officials from most NATO countries, meeting more regularly than the Ministers above them, well away from the glare of publicity, to formulate NATO's nuclear weapons needs in the 1980s and beyond. The HLG framed the two key NATO nuclear decisions taken since 1977 – the Euro-Missile decision and the Montebello decision. Ministers in the Nuclear Planning Group are politically superior, and the meetings of the NPG have high political status, but since 1977 almost the entire technical preparatory work for the NPG has been carried out by the High-Level Group, which has been known to meet the day before NPG meetings to hammer out final details of what NPG Ministers should discuss. Not surprisingly, the HLG is commonly thought of as the 'nuclear workhorse' of NATO.

With the HLG playing a major role in setting the NPG agenda, and in determining the policies for Ministers to approve, the role and status of the Nuclear Planning Directorate staff in Brussels has somewhat diminished. Where, previously, national Defence Ministries liaised via Brussels staff, they now tend to liaise directly with each other within the High-Level Group, by-passing the Brussels apparatus completely. It is widely acknowledged that the members of the High-Level Group are in a better position anyway than Brussels staff to know what issues need to be discussed, to be aware of priorities, and to know what types of proposals are worth considering. HLG members are high in the national policy-making apparatus, more in touch with daily concerns of defence decision-making than are the Brussels personnel who may be somewhat remote from them.[3]

THE SPECIAL CONSULTATIVE GROUP

The NATO Special Consultative Group (SCG) is often regarded as the sister group to the High-Level Group. It too was set up to deal with the prominent nuclear issues of the late 1970s. It comprises senior Foreign Ministry officials from national capitals, and its mandate is to review and consult on the nuclear arms control policies of NATO. Unlike the HLG, the SCG has a high public profile and communiques are usually issued after its meetings. The group is chaired by a senior official in the US State Department.

NATO SENIOR COMMITTEES

Since 1949 the NATO Council has established a large number of subordinate committees which service the Council and frame policies which the Council or the Defence Planning Committee can then adopt. These Committees, of which there are currently nineteen senior principal ones and hundreds more beneath, cover the whole range of NATO activities but, unlike the Nuclear Planning Group, they do not meet at Ministerial rank. They are generally open to all members of the Alliance, and there is broad national representation on most of them. NATO Council and the DPC cannot approve policies unless they have met with approval and endorsement at lower levels. If proposals cannot gain consensual support at the level of the relevant committee, their discussion in open sessions of the Council of DPC are liable to lead to discord or rancour – and will tend to be avoided.

> 'I bear solemn witness to the fact that NATO Heads of State and of Government meet only to go through the tedious motions of reading speeches drafted by others, with the principal objective of not rocking the boat . . .' said Pierre Trudeau, former Canadian Prime Minister. 'At NATO high-level meetings, any attempt to start a discussion or to question the meaning of the communique – also drafted by others long before the meeting began – is met with stony embarrassment or strong opposition . . .'.[4]

Military Organisation

NATO MILITARY COMMITTEE

The military task of NATO is to prepare the joint defence of the allied nations. The highest military authority is the NATO Military Committee, which has a large military bureaucracy to execute its instructions.

The Military Committee is composed of the Chiefs of Staff of each member nation (except France, which withdrew from the Integrated Military Structure in 1966, and Iceland, which has no armed forces) and meets twice yearly at Chief of Staff level. The function of the Military Committee is to recommend to the civilian Council and Defence Planning Committee those measures it considers necessary for the common defence of the NATO area.

THE INTEGRATED MILITARY STRUCTURE

The task of planning military operations is that of the NATO military commanders and their staffs inside the Nato Integrated Military Structure. In 1950–51 the NATO Council endorsed the idea of an Integrated Military Structure for all the armed forces assigned to missions in the NATO area. Until that time, the armed forces of member states had been commanded and organised on a purely national basis; and NATO's military profile amounted to no more than a US military guarantee via its nuclear superiority and its regular forces in Europe to its West European allies.

Following the vital 1950–51 decisions the Alliance military structure was transformed.[5] Today NATO's command structure and war-time planning are fully integrated between member states. NATO's geographic area is divided into four military command regions:

Allied Command Europe (ACE)
Allied Command Atlantic (ACLANT)
Allied Command Channel (ACCHAN)
The US Canadian Regional Planning Group

A supreme commander is appointed to each of the three main commands, and a subordinate hierarchy of regional military commands has been created. These commanders would be in wartime control of the military forces of all those countries having NATO-assigned responsibilities in their respective regions.

The supreme NATO commanders have large standing staffs under them; in the realm of military planning they are highly influential. Most notable is the staff under the Supreme Allied Commander Europe (SACEUR) at SHAPE in Belgium. SHAPE's staff includes three major nuclear weapons sections: Nuclear Operations, Nuclear Policy and Nuclear Concepts. The NATO military commanders have a high public profile, too: without question SACEUR – Gen. Bernard Rogers since 1979 – is the most well-known soldier on Continental Europe.

In peacetime most of the military forces of Alliance countries are under national command. American troops, Dutch air force pilots, Italian marines, British naval captains, continue to be commanded by their national superiors. The function of the NATO commands within the Integrated Military Structure is to prepare for military operations in war. This involves:

(a) Preparation and finalising of defence plans;

(b) joint and combined peace-time training exercises on a periodic basis;
(c) recommendations as to the training, equipment and allocation of forces necessary to meet the requirements of war.

In wartime channels of command would switch from national channels to NATO channels.

Bureaucratic Organisation

Most of the civilian and military committees mentioned above only meet on a periodic, often irregular, basis. They all require, however, a high degree of full-time staff support in order to execute their duties and perform their routine business. This support is provided by the following:

The NATO International Staff/Secretariat. This is an international civil service based at NATO headquarters in Brussels, numbering thousands of personnel. Its function is to support and administer the work of the NATO committees in their consultation and deliberation on policy.

The International Staff is headed by the NATO Secretary-General, the figure ultimately responsible for the effective operation of the NATO bureaucracy. He is not a politician but a civil servant: he represents no government so is not empowered to undertake diplomatic ventures, say, to the Warsaw Pact. Within NATO he may arbitrate inter-Alliance disputes; he can conciliate, initiate ideas, and will normally maintain a prominent public profile as an Alliance figurehead. He serves in many capacities within the NATO committee system.

The International Military Staff. This Brussels-based military bureaucracy numbering today some 150 officers, 150 other enlisted personnel, and 100 civilians, provides staff support to the Military Committee. Its departments – Intelligence, Plans and Policy, Operations, Management and Logistics, Command and Control, and Armaments Standardisation – cover the whole spectrum of NATO military activity, including nuclear planning. Within these divisions Military Committee policies are framed and implemented.

Civilian and Military Agencies/Institutions. In 1982 there were seventeen major NATO civilian and military institutions providing permanent specialised support and advice on NATO planning. For instance, large multi-national agencies exist in the Netherlands to manage the NATO Airborne Early Warning System (AWACS), in Munich to supervise the joint Anglo-German-Italian Tornado aircraft programme, and in Brussels to supervise integrated civil-military communications projects. These and other institutions are responsible to their relevant NATO senior civilian committees. On the military side there are agencies for NATO naval communications, advanced aviation research, equipment standardisation and many more. The NATO Defence College in Rome trains key officials. All these are directly under the auspices of the NATO Military Committee.

The North Atlantic Assembly is the inter-parliamentary organisation of member countries of the Alliance. It provides a forum where elected representatives in domestic assemblies can gather to debate and ponder NATO-related concerns.

> 'Within NATO itself there is very limited internal communication, and the management process is antiquated . . . There is no way to avoid duplication with previous studies conducted on resources because the studies are all indexed and filed by hand. There is no computerised system and no retrieval system. In fact, there is only one outdated word processor in this entire division and use has to be reserved weeks in advance . . .'[6]

NATO AND ALLIANCE POLICY-MAKING – THE PRINCIPLE OF CONSENSUS

NATO decision-making is a process in which national governments and armed forces feed their policy preferences into the collective planning and consultation process. This process is enshrined in the doctrine of NATO 'consensus'. NATO senior committees operate on the basis that decisions have to be endorsed formally, not merely by majority, but by consensus, so that NATO decisions do not come into conflict with national views. The function of NATO consultations is to negotiate between national representatives at various levels towards a position which attracts consensual support and where the problem of national–NATO conflict should not arise. By the time NATO adopts policies in the Council or DPC, the search for consensus should already have been completed. If NATO policies could be authorised by majority, immense problems would arise for dissenting countries, and NATO decisions would lose all credibility and meaning.

The theory and the reality of 'consensus' politics are not the same. 'Positive consensus' within NATO groups is best; but decisions may be based on 'silent consensus', when countries with reservations remain silent during the final phases of deliberation and endorsement.

Political pressures to stay silent, or the dangers of losing face with allies, may persuade countries to keep doubts to themselves. If strong dissenting views are maintained right up to the level of Council and DPC these may be referred to in the footnotes of NATO communiques. But the dissenting texts themselves are not readily accessible. In recent times Greece, Spain and Denmark have been notable for their frequent expressions of formal dissent from the consensus position.

Realities of the Policy-Making Process

NATO's organisational layout gives the impression of awesome formal complexity. The interrelation of NATO civil and military authorities, of national ministries and NATO delegations, of administrative and advisory authorities, seems labyrinthine. But the institutional skeleton of NATO conceals as much as it reveals. Policy-making is the outcome of many more factors than are apparent from NATO's formal machinery.

For example, permanent delegations at NATO headquarters vary enormously in size, expertise, experience and consequent influence. Some NATO military agencies perform duplicate functions, and the weight of their influence may depend on factors like scientific prowess, prestige and connections. For instance, the SHAPE Technical Centre researches many of the same questions as do the divisions of the International Military Staff, but the former, being manned largely by high-ranking Americans, and answerable directly to the prestigious SACEUR, is likely to be more highly regarded by many European NATO officials. It is widely recognised that SACEUR's voice is more authoritative than that of the Chairman of the Military Committee, even though the latter is technically senior in the NATO hierarchy.[7]

Many NATO agencies and groups, set up on an *ad hoc* basis, assume permanent existence; others function effectively even though their precise status in the NATO policy hierarchy is not clear. On most occasions critical inputs into NATO policy-framing come from expert defence and foreign ministry officials based in Washington, Bonn or London. New NATO consultative institutions have sometimes been established to reflect and enhance the dominance of national over NATO-centralised expertise. This has been the case in nuclear planning, with the High-Level Group devolving policy-making influence on sensitive nuclear issues away from Brussels. Often bilateral consultations between NATO members have a great influence, outside the multi-lateral NATO framework: many ultimate NATO decisions – on weapons development, nuclear arms control, procurement co-operation and so on – have been moulded through bilateral exchanges side-stepping the conventional NATO machinery. The Brussels bureaucracy has a reputation for being extremely cumbersome: bilateral or multilateral diplomacy outside NATO's walls can lead to speedier, more efficient and effective decision-making consultation.

NATO AS A NUCLEAR ALLIANCE

NATO is a nuclear alliance in which nuclear weapons are thoroughly integrated into the military structure and political profile. In peacetime these weapons are at the heart of NATO's strategic doctrine. In wartime they would come under the control of the NATO supreme military commanders and their subordinates. Two countries provide NATO's nuclear weapon systems – the United States, and Britain. However, the status of their forces in NATO is not as clear as the formal declarations on this matter suggest. In addition, many more NATO countries than just these are involved in NATO nuclear operational planning. To understand NATO nuclear decision-making it is necessary first to appreciate the range of NATO nuclear activity.

The United States

Europe is a major military command region in the global American security apparatus. The Supreme Allied Commander Europe (SACEUR), the American general who would command all NATO forces assigned to him in wartime, is a senior member of the American command structure.[8]

Since the arrival in West Germany of the first American Army atomic cannons in 1953, the United States has deployed many thousands of nuclear weapons in the European theatre. Numbers are rarely static: the 1984 figure was around 6000 warheads. The US Army in Europe is equipped with hundreds of Atomic Demolition Munitions, small, portable nuclear explosives some of which could be used in wartime by elite commando units to destroy enemy bridges, means of communication and other 'soft' targets. A decision was taken in March 1985 by the NPG that these nuclear land mines should be withdrawn but it was later reported that a new atomic demolition munition might be under consideration. The Army possesses hundreds of missiles and over a thousand short-range artillery shells – so-called 'battle-field nuclear weapons' – deployed in forward positions in Western Europe. These systems are being 'modernised', in other words, upgraded. And in December 1983 the US Army in West Germany began fielding the Pershing II ballistic nuclear missile, a long-range 'fast-flier' capable of striking Soviet targets within ten minutes of launch.

The US Navy in Europe owns air-dropped nuclear depth bombs, nuclear anti-submarine rockets and nuclear surface-to-air missiles for destroying enemy submarines and aircraft. American aircraft carriers in Europe carry US naval nuclear-capable fighter-bombers and anti-submarine aircraft. In addition, SACEUR also targets several Poseidon nuclear missile submarines allocated to NATO from US central strategic forces. These missiles are capable of long-range attacks on Soviet targets. The US Air Force has deployed hundreds of nuclear-capable fighter-bombers on European territory, and controls the ground-launched pilotless nuclear cruise missiles whose deployment began in Britain and Italy in 1983–4.

From the American official perspective, these weapons are theatre nuclear systems, in the sense of being earmarked for the defence of NATO and subject in wartime to the political control of NATO authorities. Decisions covering these forces – size, control, effective communication, basing, targetting, mission and so on – have been as diverse in their formulation as the weapons themselves. On the one hand, as part of the American military apparatus, decisions covering these forces can only be implemented through the American policy process. It is an American national responsibility to deploy or to withdraw a particular weapon. However, in the past decade the process of decision covering American nuclear deployments in Europe has been a matter of consultation with European NATO nations too. The two major recent NATO nuclear weapons decisions – the 'Dual Track' decision in 1979 and the 'Montebello' decision of 1983 – were reached after lengthy processes of consultation which had been non-existent in decision-making in past decades. There are strong signs that NATO's nuclear consultation institutions covering American nuclear forces in Europe will continue to function in ways signifying a historic break with the practices of the first thirty years of NATO.

British nuclear forces are described in Chapter 3.

French Nuclear Forces

In one of the watersheds in NATO history, France withdrew from NATO's integrated military structure in 1966, causing great diplomatic and military upheavals in the Alliance. Today French military forces are organised independently of NATO, and decisions on force planning are made solely within the French policy-making apparatus. French nuclear planning – covering the French nuclear force of submarines, bombers and missiles – is fully independent of NATO.

However, this is just the formal position. Various forms of military operational co-operation do go on between the French and NATO. This could hardly be otherwise given France's strategic situation. France is still represented on 345 of NATO's 380 committees.[9] In the nuclear realm the contents of this co-operation are shrouded in secrecy. To convey the flavour of this co-operation, consider the following statement of SACEUR in 1983.

> On France's role in the Alliance three principles are maintained: France is not going to rejoin the military alliance, France is keeping its independent nuclear force, and France retains the independent use of this force. But some thought is currently being given to the question as to how, if in the event of conflict France were to decide to unite with the Allied forces, *the least possible time could be lost in making its nuclear force available to the Allies.* Enormous progress has been made in this connection. *If you knew what I know you would be very encouraged . . .*[10]

Other NATO Nations

The military forces of Belgium, Greece, Italy, the Netherlands, Turkey and West Germany all have nuclear weapon responsibilities in NATO military planning. Units of the Belgian, German, Italian, Greek, Dutch and Turkish armies are trained to either fire nuclear artillery or LANCE battlefield nuclear missiles, or both. Squadrons in the German, Greek, Dutch, Italian and Turkish air forces are all capable of flying nuclear strike missions in nuclear capable fighter-bombers. The German and Italian naval air forces can deploy nuclear depth charges on their aircraft for destroying enemy submarines. All these nations' servicemen receive education in nuclear doctrine, tactics and operational procedures.[11]

Given this degree of NATO-related nuclear weapons activity, how can it be said that these countries are 'non-nuclear'? The answer is clear. None of these countries is a manufacturer of nuclear wearheads. In peace-time all the nuclear warheads which would be fired in war by Germans or Greeks, Italians or Turks, are stored in nuclear warhead stockpiles held in custody by US personnel. Only once an American decision is taken, at a time of crisis, alert or war, to disperse these warheads from their sites do the military forces of the other allies 'go nuclear'. The American warhead sites are officially referred to as 'special ammunition sites'; their locations are never revealed, for obvious reasons (though it is clear they are to be found in over 100 strategically vital sites); nor are the procedures for nuclear dispersal known.

The existence of these arrangements means that, in instances, nuclear roles are, literally, shared in NATO wartime operational plans. America provides the warheads, allies provide the delivery systems – the aircraft, missiles, artillery and so on. These arrangements are (loosely) referred to as 'dual-key' arrangements.

Given these trans-national sharing responsibilities, the distinctions between nuclear and non-nuclear nations become fuzzy. Once a state of serious crisis alert has been reached, NATO has procedures for the large-scale dispersal of nuclear warheads from their storage sites. If this happens the military forces of seven European countries could gain access to nuclear warheads. In such circumstances neither the electric 'locks' on the weapons (known as PALS) nor the rigid political procedures for authorising nuclear weapons use, would prevent the possibility of unauthorised nuclear attacks by, say, Italian, Turkish or Greek fighter pilots in the midst of a crisis atmosphere or conventional warfare.[12] This illustrates the meaning of NATO as a nuclear alliance, not only in the political sense, but in the military operational sense, with several countries' armed forces, in certain scenarios, capable of initiating nuclear operations. It points also to the fragility and ambiguity of NATO nuclear command processes.

Basing of US Nuclear Warheads and Weapons in Nato-Europe

When the United States began basing US nuclear systems permanently on foreign soil in the 1950s it concluded a series of bilateral agreements with the host nations concerning nuclear deployments. These are known as *Programmes of Co-operation* (POCs): they permit the transfer of US nuclear weapon delivery vehicles onto foreign soil; they permit the permanent siting of US nuclear warheads on the territory of the state concerned; and they may also cover training in the United States of units of the host nations' armed forces involved in the operation of the nuclear weapon systems provided to them. In 1983 POCs existed between the United States and eight NATO countries: Belgium, Canada, Greece, Italy, the Netherlands, Turkey, Great Britain and West Germany. A POC agreement also exists with South Korea. The US has nuclear warheads and delivery systems in all these countries.[13]

The decisions to establish these Programmes of Co-operation were generally taken in secret, and their precise contents are secret today.

POCs provide the legal framework for the siting of nuclear weapons overseas. They are highly sensitive, and any break in a POC is certain to create tension. POCs are intended to uphold the principle that US weapons are sited overseas with the 'consent' of the nation concerned. In reality, the consent is that of a very few defence and foreign ministry officials and senior government personnel.

Periodically, evidence is produced that the diplomatic courtesies of POC agreements have not always been adhered to. In February 1985 secret Pentagon documents were leaked, revealing that America had secret plans to station nuclear weapons in foreign countries in an emergency without seeking prior permission. The countries include Canada, Spain and Iceland (all in NATO), Bermuda, Puerto Rico and the British Island of Diego Garcia in the Indian Ocean. These plans are not covered by any standing POC agreement; the Pentagon plan stated explicitly that none of the countries would be informed.[14]

Disclosure of the plans created diplomatic storms, which US officials moved quickly to quell. A Presidential aide said: 'We are in touch with all the involved Governments and explaining just what we intend . . . We may have made a mistake in not doing so earlier . . .'.[15] But Richard Perle, a senior official at the Pentagon (and chair of the NATO High-Level Group) was unrepentant: 'I don't see how in the long run we can ask the American people to bear the risks of war to defend allies who will have nothing to do with us when issues like the movement of nuclear weapons is concerned . . .'[16]

Episodes like this raise the question whether POCs would actually matter at a time of military crisis, and whether all that POCs really do is provide the peacetime trappings of democratic legitimacy to secretive wartime plans and deployments.

Nuclear Decisions: the Case of Holland

It is worth looking briefly at one example of the extent of nuclear decisions which have to be taken by any NATO ally which has chosen to have nuclear missions and roles inside the integrated military structure. In the case of Holland, for example, it is generally believed that governmental nuclear decision-making covers NATO-wide programmes or US initiatives, for example the deployment of cruise missiles or the Enhanced Radiation Warhead (the 'neutron bomb'). In fact, the ambit of Dutch nuclear decision-making is wider.

Currently Holland has several nuclear weapon responsibilities within NATO's military framework. For instance, Dutch Air Force squadrons still man the NIKE HERCULES nuclear air-defence system; and army batteries are trained in the use of LANCE missiles. It is a decision of Dutch political authorities to add, subtract or modify nuclear weapons roles for its armed forces. In NATO, 'non-nuclear' countries such as Holland make nuclear decisions all the time.[17] The distinction between non-nuclear and nuclear powers confuses as much as it clarifies in NATO's case.

Not all the decisions taken by 'non-nuclear' members of NATO are positive decisions endorsing weapons responsibilities. Some of the most well-known have been decisions to refuse some activity. For example: in the 1950s Norway, Iceland and Denmark decided not to host any nuclear warheads on their territory in peacetime; and successive West German governments have forbidden the German armed forces access to any nuclear weapon system – bomber or missile – capable of striking Soviet territory.

Command and Control of NATO Nuclear Forces

Command and Control (C2) systems provide the means by which to communicate with nuclear weapon units of the armed forces, and to direct their use. These systems are the 'central nervous systems' of the nuclear warfare establishment: without them the weapons are inert bits of technology, their military custodians unable to receive commands about what to do: with them the weapons can fulfil the missions for which they are intended, and they can serve – in theory – the ends which the political leadership desires in war.

The programmes involved here are complex, expensive and technically formidable, but they are largely away from the political limelight. C2 systems are the hidden, unglamorous side of the nuclear establishment. Few politicians, or members of the public, choose to understand the technologies involved, or their precise purposes. Many of the most important programmes are funded over many years, and are concealed deep within military requests full of acronyms and weird technical terms; in Europe, if not in the US, C2 has a low budgetary and political profile. But the fact remains that C2 is at least as significant to nuclear deterrence and to nuclear warfare as specific weapons systems.

NATO has several agencies devoted to technical work on C2 problems. This involves research and development into communication technologies; advanced satellite surveillance, radar systems, electronic warfare counter-measures, the provision of airborne command posts for senior military commanders, and much else. These agencies include: the NATO Integrated Communications System Organisation (NICSO) in Brussels, and several specialised military agencies responsible to the NATO Military Committee. The International Military Staff at Brussels includes a C2 division; and several senior civilian committees assess C2 problems and proposals. The staff under SACEUR at Supreme Headquarters Allied Powers Europe (SHAPE) includes a Nuclear Command and Control division of high-level military and scientific experts. As in other areas of nuclear activity, these SHAPE personnel probably carry most weight in nuclear decision-making in this field.

NATO's plans for using nuclear weapons in war are formulated by the nuclear weapons sections of SHAPE. The staff in these operate under NATO guidelines contained in a document called 'Concepts for the Role of Theatre Nuclear Strike Forces in Allied Command Europe' adopted by the North Atlantic Council in 1970. These political guidelines have been supplemented by others, but the precise operational details of targetting and nuclear use options are very much in the hands of military planners. Nuclear war plans are rarely static. As new weapons are procured, and old ones retired, war plans must be altered to accommodate the changes. Sometimes weapons have been procured before any clear roles in war have been conceived for them. At other times new target lists have been drawn up without the weapons available to hit them. In short, it is a mistake to think that war planning and weapons procurement have been, or are, tightly co-ordinated.[18]

NATO's basic war plan covering longer-range nuclear weapon systems is the Nuclear Operations Plan developed by SACEUR for the execution of nuclear strikes with the weapons assigned to his command. These include land-based nuclear-capable aircraft; land-based intermediate-range missiles; carrier-based nuclear-capable aircraft and the submarine-launched ballistic missiles which the US and UK have assigned to NATO. NATO's short-range nuclear systems are covered by a different set of war action plans.[19]

Consultation and Participation

NATO's nuclear arsenal is not a homogenous entity. NATO's nuclear forces comprise the sum of national nuclear forces specifically assigned to NATO, and in addition the essential support structures such as command centres and communications systems which make these nuclear forces into viable fighting units. Many of the policies which govern these forces are framed, at least in part, in NATO-wide consultative groups. But others never reach NATO chambers of consultation and stay firmly in the national policy-making apparatus. Only national governments can implement nuclear decisions once reached. NATO-wide consultative groups are not executive bodies. Their business is diplomacy, technical deliberation, sharing of information and viewpoint. As the United States has the dominant nuclear role in NATO it is not surprising that the United States dominates the consultative process. But once the High-Level Group, the Nuclear Planning Group or the Defence Planning Committee itself reaches a decision on nuclear matters, it is for the country or countries concerned to execute its part of the decision.

In sum, the existence of NATO's present nuclear consultation process means that more countries participate in nuclear deliberations on nuclear policy, prior to decision, than ever before. The decisions, once made, are thus legitimated as NATO decisions, not merely decisions made 'for NATO' by the United States. The question is whether these NATO consultative institutions – notably the Nuclear Planning Group and its offspring the High-Level Group – have actually resulted in wider NATO participation and confidence in US nuclear policy as it affects Europe, or whether these groups provide the appearance of European-wide participation in policy-framing when, in fact, Washington still controls the actual process.

> 'Decisions about NATO deployments must be made, and have been made, even though definitional problems remain. In the practical world, the mere lack of an unambiguous set of objectives is hardly an obstacle to the acquisition and deployment of thousands of nuclear weapons. Defense organisations have a way of running themselves even without clear and consistent guidance. Choices have to be made, and they are made. A failure to define unambiguous centralised direction for a defense system means only that direction is supplied from decentralised, bureaucratic sources . . .'.[20]

This chapter has examined, albeit briefly, the structure of NATO and how the decision-making process works in theory, including the principle of consensus: the realities of the basing of nuclear weapons in NATO countries has been discussed, and their control; it has become clear that the process of consultation on nuclear issues has been extended. A case study follows, which looks in detail at the process by which a major NATO nuclear decision was reached.

THE EURO-MISSILE DECISION 1979 – A CASE STUDY

In December 1979 NATO decided to deploy new American Pershing II and cruise missiles on the European continent, beginning in late 1983. The North Atlantic Council, the Defence Planning Committee and a Special Ministerial session all issued final statements of decision, which have taken on an importance few could have foreseen at the time. The decision has almost legendary status today, and its political implications have been huge.[21]

Intellectual Roots of the Decision: 1974–77

The 1979 decision did not happen out of the blue. It was preceded by years of expert analysis and political consultation within NATO. Very significant in the early stages were the attempts to find intellectual rationales and justifications for 'upgrading' NATO nuclear posture.

Since the 1950s experts have been preoccupied with America's nuclear commitment to Europe. The dilemma of how America can provide nuclear 'protection' for Europeans has always generated political and intellectual heat, and the decision to place Euro-missiles in Europe needs to be seen as the latest in a historic series of responses to this dilemma.

There is no single moment when the ideas which led up to the 1979 decision took root. By 1975 defence experts on both sides of the Atlantic had begun to devote attention to American nuclear deficiencies and weapons gaps in Europe. But few politicians were aware of this issue at all. Expert discussion emanated from the International Institute for Strategic Studies (IISS) in London, and from the Institute for Foreign Policy Analysis in Philadelphia, USA. In 1976 a special think-tank of defence analysts was convened to assess in greater detail the requirements of NATO's nuclear profile. The group was called the Euro-American Workshop, and it met over a period of two years. The lobbying efforts of this group, and their collective intellectual weight, proved highly influential.

No politician belonged to the group, but political figures often attended seminars, and the group had the ear of high-level ministry officials and NATO staff. It was this group which, in its papers, briefings and seminars, shaped the concepts, the justifications and the strategic explanations, for Euro-missile deployment.

Defence Ministry Interest: America and Europe 1974–77

Parallel to this intellectual interest in new nuclear systems for Europe came a degree of official governmental activity. As early as 1975 a Pentagon report to Congress declared that US theatre nuclear weapons in Europe should be modernised. But the report did not go into specifics. European Defence Ministry personnel were also paying increased attention to nuclear systems in Europe, and by the summer of 1977 interest in new American missile technologies – notably the cruise missile – peaked. European officials saw cruise missiles as a cheap, versatile weapons system of the future which might be useful in their own upgraded arsenals – conventional or nuclear – and which, in the hands of the Americans, could provide renewed 'guarantees' to Europe in an era of the balance of terror. By Autumn 1977 defence ministry officials on both sides of the Atlantic were seriously contemplating new nuclear options for NATO. But senior ministers, heads of states, were not yet showing interest or concern about the matter.

NATO Organisational Activity: the Creation of the High-Level Group

The bureaucratic momentum in NATO behind the decision began to build up in 1977. In May, fifteen NATO heads of government met in Washington DC and endorsed a wide-ranging programme of defence improvements. One of the ten general areas for improvement was designated as NATO's theatre nuclear forces. To examine this in detail, and to formulate specific recommendations, NATO's ministerial Nuclear Planning Group set up, in October 1977 and at the strong urging of the United States, an expert sub-committee of defence ministry officials from eleven countries which became known as the 'High-Level Group'. In the next two years, the High-Level Group, its work dominated by the United States, was to frame the Euro-missile proposal.

Helmut Schmidt Speech 1977: the Political Trigger of the Decision

In late October 1977 came the first serious hint from a NATO political leader of concern about NATO's nuclear strategy. In his speech to IISS (which mostly concerned global energy and resources issues), West German Chancellor Helmut Schmidt included a passage which was interpreted as a warning that NATO's military profile was defective. He spoke of regional nuclear imbalances in Europe, of the harmful effects of detente and SALT on the nuclear situation on the continent.

This speech is regarded almost universally as the European 'call' to the Americans to come to Europe's aid with some new nuclear commitment. But Schmidt's speech was not a surprise to everyone, nor was it an original call. Defence ministry officials in the Pentagon and in major European capitals knew just what Schmidt was talking about; so did the defence analysts in the Euro-American Workshop who were only too keen to spread word of these matters; what is more, NATO had already set up a new group – the High-Level Group – mandated to respond to the challenge Schmidt had conveniently laid down. By the time Schmidt spoke up, there were already a varied set of officials and lobby groups keen to interpret Schmidt's words in ways which suited their bureaucratic and intellectual preferences, which had taken root long before Schmidt uttered a word. Schmidt's speech is significant because it is a *quasi*-mythical point of reference in all political discussion of the decision-making process. But the reality is that Schmidt did not invent some new argument: he voiced the arguments of others, and whetted their appetites to see through the logic of his words.

Political Uncertainties in the USA 1977–78

The High-Level Group began meeting in November 1977. Although it was attended by defence ministry personnel from ten NATO–Europe capitals, none of them had the resources or the knowledge to match the American impact in the group.

By February 1978 the High-Level Group had concluded that new American systems were required for Europe, if NATO's nuclear dilemmas were to be resolved. But their assessment was technical and obscure, and it is clear that senior American politicians, especially the 'top two' in the US State Department, Cyrus Vance and Les Gelb, were not convinced that the HLG conclusion was correct. At the time the view of the State Department seemed to be that Pentagon officials were aggressively driving forward a policy option in the High-Level Group without anyone properly thinking through the diplomatic and political repercussions.

The Neutron Bomb Fiasco

In April 1978 this mood of uncertainty dramatically changed. NATO suffered acute embarrassment over the US Army's forthcoming 'Neutron Bomb' deployment which President Carter had spent a year persuading his European allies to 'ask for' but which, without explanation, he had then decided to postpone. Carter was condemned by the Germans in particular for pulling the rug from under their feet; he was criticised for indecision and weakness in the face of public and Soviet protests; and his entire political reputation was tarnished.

This episode had a cathartic effect on the American leadership. What was needed now to repair the damage was resolve, leadership, determination and decisiveness in Washington. Only then could Europe receive the diplomatic reassurance it needed after the neutron bomb fiasco. From this time on, the work of the High-Level Group assumed an importance which was no longer merely technical or military – it became political, too. American doubters of 'modernisation' – such as Vance and Gelb – rapidly became converts when they realised the political necessity to 'see the decision through'.

In June 1978 the US President mandated in a directive known as PRM-38 a top inter-departmental study of NATO nuclear modernisation, the objective being to reach a powerful consensus behind the programme in the American policy-making apparatus.

In October the PRM-38 group passed a set of 'military options' papers, prepared for it by the Pentagon, to the High-Level Group. Two of the four options outlined were new missiles known as the Pershing II and the Tomahawk cruise missile. The High-Level Group went away to assess in detail the make-up of the weapons package required, and to fix upon a specific recommendation. During this time the Nuclear Planning Group and Defence Planning Committee each reviewed HLG work regularly. Ministers were kept informed of progress on the studies; but almost the entire input to HLG work came from Pentagon staff papers, and the important exchanges were those taking place in the high echelons of the American establishment.

The Guadeloupe Summit: the Decision Crystallises

It was at this stage that the American President himself was brought in to the decision-making process. He was urged by the National Security Council to seek the support of major European leaders for the decision as soon as possible, to ensure that there would be no repetition of the neutron bomb affair. This Carter did. In January 1979 he met with Callaghan of Britain, Schmidt of West Germany and Giscard d'Estaing of France, on the Caribbean island of Guadeloupe. All four men agreed to the principle of nuclear modernisation. This was the political signal the Americans needed. They now had the technical modernisation package, being finalised under the auspices of the High-Level Group; and they had the backing of major European politicians. From then on American diplomacy intensified behind the evolving decision. In February–March 1979 high-level US missions were dispatched to European capitals to convince sceptics to support the plan: the Germans and the British were generally compliant; the Dutch and Belgians nervous, and the Italians keen to get involved, largely, it seems, as a matter of prestige, and because they had resented not being invited to Guadeloupe in January. 1979 became the year of American positive diplomacy, in which the Americans sought, with a great deal of energy and commitment, to 'engineer' total political unity in NATO-Europe prior to the announcement of the decision itself.[22]

European Political Support

Two developments made the American task easier. First, in April 1979, they agreed to the creation of the so-called Special Group (reconstituted in 1980 as the Special Consultative Group) to review the nuclear arms control position of NATO. Its task was to set out guidelines for future arms control talks to proceed parallel to nuclear missile deployments. European leaders – especially the Germans and the Dutch – found this US interest in arms control most helpful in defusing public opposition to the evolving missile plans. With this arms control pledge behind them, the Americans were in a stronger position for their missile diplomacy to succeed.

Secondly, in 1979 concern peaked in Europe about new Soviet weapons, notably the SS-20 ballistic missile aimed at Europe. This missile became a symbol of the Soviet threat. Although the SS-20s existence had only little to do with the detailed rationale for US Euromissiles, its prominence in the news media, in political speech-making and in NATO Committee discussions, helped to create the emotional climate in which US missiles could be supported, and in which NATO unity and decisiveness came to be seen as politically vital.

Finalising the Decision: Autumn 1979

In the summer final details of the weapons package were worked out. In July senior American officials settled on a 'compromise' figure of 572 missiles for Europe – 108 Pershing IIs and 464 cruise missiles – to be based 'visibly' on land in five European countries[23]. This decision was conveyed to the High-Level Group and it soon gained consensual acceptance in the group. The HLG worked out final details of the deployment programme, and in September 1979 prepared its final report. At the same time the newer Special Group prepared its final report on NATO arms control policy. The two documents were presented to NATO defence ministers in the Nuclear Planning Group in November 1979. The NPG endorsed both documents. In the next few days officials at the US National Security Council literally stapled the two reports together into a single 'twin' document. This document is known as the IDD – the Integrated Decision Document – or, more commonly, as NATO's 'twin-track' decision. At the end of November the Permanent Representatives at NATO headquarters in Brussels approved the IDD.

The Decision is Announced

Three weeks later NATO's defence and foreign ministers collectively endorsed the IDD, whose basic contents were fully revealed for the first time. The communiques which were produced as a result of these meetings are the authentic statement of NATO's 'twin-track' decision, a decision which was the outcome of a process of technical, military and political activity stretching back over several years.

1980–85: the NATO Consensus Erodes

At the December 1979 ministerial meeting the US Secretary of State Cyrus Vance stated: 'The meeting has been one of the most productive in years. . .the decisions of 1977–9 have set the Alliance on a sound course for the 1980s. . .'[24]. Historically, these words have proved to be mistaken. Even at the time of the decision the political consensus the Americans had pursued, and which the communiques spoke of, was more fragile than outward appearances suggested. Right up to the day of the December meeting Dutch and Belgian politicans had deep misgivings about the missile programme. On the day the two countries agreed to the principle of modernisation, but deferred committing themselves to precisely timetabled missile deployments on their soil.

In 1980–83 the American programme to develop, manufacture, and test the missiles was stepped up, to meet an initial target date for deployment of December 1983. Arms control talks commenced in November 1981 between the USA and the USSR, but these collapsed two years later without success, as the American missiles began to arrive.

In this four year period the European consensus was eroded. Opposition parties in West Germany, Britain, Norway, Belgium and Holland came out against modernisation. The Greek and Danish governments registered official opposition. Public opposition was widespread and intense. The political unity in Europe behind the decision began to crumble, tarnishing the 'united' and 'resolute' image which NATO ministers, and especially the Americans, had so desired in 1979.

Analysis of the Euro-Missile Decision

The most striking error of judgement made within NATO in 1979 was to assume that the meticulous inter-governmental consultations that had gone on in NATO vested the final decision with a popular stamp of approval. 'If there ever has been an important NATO decision that was the product of true inter-governmental consensus, it was the December 1979 decision. . .' comments one observer[25]. This may have seemed true and praiseworthy at the time, but it gave rise to serious miscalculation.

The High-Level Group and the Nuclear Planning Group are secretive bodies. They have always been somewhat sealed off from everyday political issues and tensions, on the understanding that nuclear planning can be pursued in some sort of vacuum. A Congressional report on modernisation in 1981 explained the fallacy of this arrangement:

'The High-Level Group work on the strategic, military and technical considerations was extremely thorough. . .but the consultative process was restricted to the level of government officials and technical experts and did not extend to parliamentary or public opinion. . .'[26]. This was the source of NATO's error of political judgment: to assume that governmental consultation on nuclear planning is the same as public consent for nuclear plans. European citizens have proved since 1980 that this is clearly not the case. What is required to gain public support for NATO decisions are visible, democratic and authentic forms of reassurance of peaceful intent. The NPG/HLG decision-making process, greeted in 1979 as a NATO success story, created a false sense of political satisfaction, and contributed to NATO's political crisis.

> '. . .the advent of the peace movement since 1979 has gone a long way to undermining the decision-making process in the Alliance, which for all the mention of participational democracy in the Atlantic Charter has always been the responsibility of a very narrow elite. . .'[27].

Since 1979 meetings of the Nuclear Planning Group have provided opportunities for the group 'heavyweights' – USA, UK, West Germany and so on – to discourage ministers with doubts about the Euromissiles from breaking ranks on the decision. Between 1981 and 1983 ministerial attention turned to the arms control talks concerning Euromissiles: but since the missiles started to arrive, concentration has focused on the two nuclear 'weak links' – the Belgians and the Dutch – to try to ensure that they overcome their continued hesitation and take their share of the American missiles.

POLITICS OF THE NUCLEAR PLANNING GROUP

The activities of the NPG, and the preparatory work carried out in its name, have been dominated by the United States. That this is so is hardly surprising, for American nuclear policy is the subject of the vast majority of the NPG consultation. American officials alone have the expertise and the experience to guide the rest of the allies on many policy matters. In addition, much of the technical information is highly classified, and the Americans are, for reasons of history or legislative restriction, unwilling to divulge to their allies all their nuclear know-how anyway. Large chunks of nuclear information, especially on targeting, do not reach the ears of the NPG ministers.

The NPG has been an educative experience: but the educational process has been largely one way. European inputs into NPG work have tended to be limited to matters where the governments concerned have a concrete interest, for example the provision of nuclear base facilities for American weapons on national soil.

The US Secretary of Defense invariably sets the tone for ministerial meetings; he is always their central figure. Many European ministers lack interest, experience, skilled back-up or the intellectual resources required to challenge the American input into NPG deliberations. For many countries their place on the NPG is more about prestige than influence. There are only a few examples – one is Denis Healey in the 1960s – when European ministers made substantial and sustained inputs of their own into NPG work.

The 'Americanisation' of NATO nuclear consultation has intensified with the creation of the High-Level Group. Almost all the HLG staff work has emanated from the Pentagon; reliable reports of HLG meetings record a startling degree of European non-active participation in HLG preparatory work, especially prior to 1980.[28]

The conclusion to be drawn is that NATO nuclear policy is no less dominated by America than was the case in the past. The impact of the

Nuclear Planning Group has not been to cause US strategic policy to mellow or moderate; its role has been to give political figures in NATO-Europe high-level access to US policy planning. The NPG is a political forum for consultation, rather than a military decision-making group. These decisions – past, present and future –will continue to be dominated by US policy-makers and military personnel.

The real purpose of the Nuclear Planning Group has been to endow nuclear policies covering NATO with an aura of NATO-wide authenticity. All recent decisions have been 'NATO decisions on US nuclear weapons' not 'US decisions' alone.

The fact that NATO defence ministers from thirteen countries meet, twice-yearly, to review nuclear plans is a potent political symbol of the diplomacy of consultation. From the perspective of the governments concerned, the existence of the NPG is 'proof' that NATO-Europe cannot simply be the nuclear yes-men for the Americans. Citizens in NATO-Europe may have their own views of such an interpretation.

To suggest that the NPG has been a political success is to fly in the face of NATO's nuclear history. In 1977–79 the NPG, for all its participatory credentials, failed to anticipate the changing mood of European public opinion. Often, indeed, sensitive nuclear planning decisions have not been put on the NPG agenda at all. These include: the American neutron bomb decision (1977); the American decision to build a fall-back wartime command centre in High Wycombe, UK (1982); President Reagan's so-called 'Star Wars' commitment (1983); and many others. All these decisions arrived on the NPG agenda after they had burst on to the public scene, and after the political controversy – in other words, when it was too late. Such examples suggest that the Nuclear Planning Group as an institution is unresponsive to public attitudes on NATO nuclear affairs.

NOTES

1. The basic reference book on NATO, which includes summaries of major North Atlantic Council decisions, is *The North Atlantic Treaty Organisation: Facts and Figures* (Brussels: NATO Information Service, 1984).
2. A useful recent study of the Nuclear Planning Group is by P. Buteux, *Nuclear Consultation in NATO: 1965–1980* (Cambridge University Press: 1983).
3. For a discussion of this point, see D. Schwartz, *NATO's Nuclear Dilemmas* (Brookings: 1983) pp. 244–5.
4. Speech on receiving the 1984 Einstein Peace Prize; reported in *Guardian* (15 Nov. 1984).

5. See *NATO: Facts and Figures*, op. cit., esp pp. 27–28.
6. Robin Beard, Assistant Secretary-General for Defence Support, interviewed in *Aviation Week and Space Technology* (21 May 1984).
7. See J. Eberle 'Defence Organisation in the Future' ch. 6 of L. Martin (ed.) *The Management of Defence* (London: Macmillan, 1976).
8. For which, see D. Charles 'Who Controls NATO's Nuclear Weapons?', in *Bulletin of the Atomic Scientists* (Apr. 1985) esp. p. 48.
9. From 'Reshaping the Alliance' special report in *Aviation Week and Space Technology* (21 May 1984).
10. Quoted in *Atlantic News* (4 Feb. 1983).
11. For full details on European NATO nuclear-capable systems see, 'Background Materials on Tactical Nuclear Weapons Primarily in the European Context' in SIPRI's *Tactical Nuclear Weapons: European Perspectives* (London: Taylor & Francis, 1978); *The Military Balance 1984–1985* (London: IISS, 1985) pp. 131–2 and P. Bracken *The Command and Control of Nuclear Forces* (Yale University Press, 1983) ch. 5.
12. For a remarkably penetrating analysis of NATO nuclear doctrine see P. Bracken, op. cit., ch. 5, esp. pp. 164–78.
13. See T. Cochran et al, *Nuclear Weapons Databook: vol 1 – US Nuclear Forces and Capabilities* (Ballinger, 1983) p. 94.
14. *Washington Post* (14 Feb. 1985).
15. *Daily Express* (15 Feb. 1985).
16. *Guardian* (16 Feb. 1985).
17. See The Hague, Ministry of Defence, *The Netherlands Defence White Paper 1984: Summary and Excerpts*, pp. 6–11.
18. See D. Ball 'Strategic Forces: How Would They Be Used?' *International Security* (Winter 1982–83) esp. p. 47; on the relationship between war plans and nuclear weapons procurement see J. Richelson, 'The Reagan Strategic Modernisation Programme', *Journal of Strategic Studies* (June 1983).
19. For a revealing and authoritative presentation of NATO war-plans see D. Ball, *Targeting for Strategic Deterrence* (IISS: 1983), pp.19–21.
20. P. Bracken, op. cit., p. 135.
21. The literature on the decision is extensive. The following account relies particularly on R. Garthoff, 'The NATO Decision on Theatre Nuclear Forces', *Political Science Quarterly* (Summer 1983) pp. 198–212; relevant chapter in R. Betts (ed.) *Cruise Missiles: Strategy, Technology, Politics* (Brookings Institution, 1981); D. Schwartz, op. cit.; P. Buteux, op. cit. and several *Congressional Research Service* papers and sections of *Congressional Testimony*.
22. The phrase is from R. Garthoff, op. cit., p. 204.
23. General Rogers told the US Senate that a sea-launched cruise option had been considered, but rejected because of lack of visibility to the European public.
24. Quoted *Atlantic News* (15 Dec. 1979).
25. From D. Schwartz, op. cit., p. 243.
26. From *The Evolution of NATO's Decision to Modernise Theatre Nuclear Weapons* (US Library of Congress Research Service, report: 1981).
27. C. Coker, *Future of the Atlantic Alliance* (Royal United Services Institute. 1984) p. 47.

28. See in particular F. Kaplan 'Warring Over New Missiles for NATO', *New York Times Magazine*, (9 Dec. 1979) and G. Treverton, 'NATO Alliance Politics', ch. 13, R. Betts (ed.), op. cit., esp. p. 427.

7 Warsaw Treaty Organisation

Studying the Warsaw Treaty Organisation (WTO) is a difficult undertaking. Our knowledge of its functioning depends partly on published Soviet and Eastern European documents and communiques, partly on information picked up by Western journalists who report on its meetings, and partly on information on its military aspects which is reported by Western intelligence sources. One recent Western commentator has noted that '. . .with an alliance such as the Warsaw Pact, whose workings proceed behind a veil of secrecy, the greatest caution, even humility, is necessary[1].'

In spite of these inherent difficulties, it is possible to make some comments on those aspects of decison-making within the WTO which are most closely related to nuclear weapons and nuclear policy. The WTO provides a contrast with NATO in that it contains only one nuclear power – there are no middle-ranking 'independent' nuclear powers like France and Britain within the WTO, and as far as is known no nuclear warheads are in the hands of any WTO forces apart from Soviet forces. There are Soviet nuclear weapons deployed on the territory of certain Eastern European states within the WTO, but control of these weapons appears to remain in Soviet hands, with no evidence of any dual-key arrangements similar to those which operate within NATO.[2]

Consequently, major decisions on the design, development, and manufacture of nuclear weapons are a prerogative of the Soviet Union, and not issues for the WTO as such. This aspect of the WTO is partly a reflection of the fact that the Soviet Union plays a more dominant role within the WTO than the USA plays within NATO, in terms of military forces and expenditure as well as nuclear weapons. In this sense, nuclear weapons decision-making procedures are less crucial to an understanding of the WTO than they are in the case of NATO. However, WTO structures do have to deal with some questions concerning nuclear deployments, their foreign policy implications, and, to a lesser extent, the financing of nuclear policy. Since WTO military doctrine and strategy are essentially determined by Soviet thinking and then adopted by the alliance as a whole, it is possible and important to look at the nature of nuclear decision-making within the WTO.[3]

This chapter provides a brief history of the WTO since its formation in 1955; a sketch of the WTO's military command structure and forces with some comments on nuclear deployments in Eastern Europe; comments on WTO political structures; and a case-study of the process of recent nuclear 'counter-deployments' in Eastern Europe, which illustrates the workings of some of these structures.

HISTORY OF THE WTO

The Warsaw Treaty was signed in Warsaw on 14 May 1955, by representatives of Albania, Bulgaria, Hungary, the German Democratic Republic (GDR), Poland, Romania, the Soviet Union and Czechoslovakia. The immediate reason for the formation of the WTO at that time was the formal accession to NATO, a few days earlier, of the Federal Republic of Germany. Before 1955, the Eastern European states were linked to each other by bilateral treaties and through the multilateral economic organisation COMECON (or, CMEA, Council for Mutual Economic Assistance).

The text of the treaty emphasises the parties' view of the threat to European security caused by West Germany's remilitarisation and reintegration into the Western bloc. It commits its signatories to mutual assistance 'in the event of an armed attack in Europe on one or several states that are signatories of the treaty', to the establishing of a joint command for their armed forces, and to working towards the prohibition of nuclear weapons and the establishing of a collective security system in Europe.[4]

In addition to the perceived threat from West Germany there were a number of other factors at work which prompted the formation of the alliance in 1955. At that time, the post-Stalin manoeuvrings for power within the Soviet leadership were unresolved, and different groups saw the alliance as important for different reasons: to some it was a vehicle for strengthening the socialist bloc, to others, more of a bargaining-chip for managing relations with the West as tentative moves were made towards improving East–West relations. An alliance was also seen as a way of managing Soviet control of Eastern Europe at a time when

Stalinist methods of direct control were having to be eased, and of managing military co-operation more efficiently than formerly. In addition, the withdrawal of Soviet forces from Austria in 1955, under the terms of the Austrian State Treaty, created problems for the legitimisation of the Soviet military presence in Hungary and Romania.[5] Soviet writers often cite Western alliances formed outside Europe (ANZUS, SEATO) as contributory factors – a good indication of the WTO's function in a global political role as well as a regional military role.[6]

Since 1955 there have been a succession of political crises in Eastern Europe which have served to illustrate the fragility of Soviet-East European political relations and the degree to which the WTO has functioned as a mechanism for internal political control. Paradoxically, however, the institutions of the WTO itself have rarely been used to resolve these crises. In 1956, the Soviet Union re-established its authority in Hungary through the use of its own troops. Although Imre Nagy announced Hungary's withdrawal from the WTO at the height of the crisis, the Soviet Union's subsequent citing of the Warsaw Treaty as a justification for its intervention was unconvincing, since the treaty clearly refers only to defence against external attack. There seem to have been bilateral consultations between the Soviet leadership and other WTO leaders, but no co-ordination of policy through WTO channels. Events in Poland at the same time were resolved without Soviet intervention.

Between 1955 and the Czechoslovakian crisis of 1968, developments took place in the military organisation of the WTO which had implications for nuclear policy within the alliance. Both Soviet and Eastern European forces were equipped with nuclear-capable short-range weapons systems (missiles and aircraft), and trained in multilateral exercises for combat in nuclear conditions. Although evidence on this period is rather sparse, it does not appear that the Eastern Europeans were themselves equipped with nuclear warheads or trained to use them. However, there is evidence that both Czechoslovakia and Romania felt that decisions taken at this time on military planning were made with insufficient consultation by the Soviet Union of its allies. It is hard to find evidence of whether or how these issues were discussed within WTO institutions, and most commentators treat it as a question of Soviet strategy. At any rate, unhappiness with aspects of WTO policy certainly emerged in Czechoslovakia and Romania, and there is evidence that Czechoslovakia resisted pressure in the early 1960s for Soviet troops and nuclear-capable missiles to be stationed in Czechoslovakia.[7]

In 1968, the force which occupied Czechoslovakia was multinational, but overwhelmingly composed of Soviet troops. Before the intervention itself, pressure was put on the Czechoslovakian leadership by a series of meetings between WTO leaders, but none of these were technically meetings of WTO bodies. Any WTO meeting would have had to involve Romania, and Romania would not have endorsed any preparations for intervention. The intervention led Albania formally to withdraw from the WTO, although Albania had been estranged from the Soviet Union since 1961.

Romanian unhappiness with the alliance had been growing since the early 1960s, and grew out of a mixture of economic grievances and discontent over Soviet domination of the military alliance. Romania refuses to allow foreign toops on its soil, does not participate in multilateral exercises, makes regular calls for mutual bloc dissolution, and has resisted pressure for increased defence spending. Although these Romanian gestures of independence may be of limited practical significance because of Romania's lack of strategic importance to the WTO, it does seem likely that the re-organisation of WTO institutions which took place after 1968 was partly prompted by Soviet concern that insufficient consultation within the alliance could lead to Eastern European unhappiness and repetitions of the Czechoslovakian crisis.

Several additional bodies were created after 1968 which gave Eastern European leaderships more opportunity to discuss military and foreign policy, although this did not amount to giving them a veto over Soviet policy. These reforms did not, however, prevent occasional public differences of opinion, such as occurred over defence spending at the 1978 meeting of the WTO's Political Consultative Committee.[8] The imposition of martial law in Poland in 1981 illustrated yet again the seriousness of internal political problems within the WTO, although it could be argued that the imposition of a 'solution' by domestic military forces was preferable to Soviet intervention; even so, the WTO clearly continues to suffer from internal political strains which are bound to affect its military aspects.

MILITARY STRUCTURES

Figure 7.1 sets out the WTO's principal institutions under the headings 'military' and 'political'. It is, perhaps, slightly misleading to suggest that there is a sharp dividing-line between the two sets of institutions, and it is in the nature of the WTO (as with other alliances) that one cannot always easily distinguish military from political concerns.

MILITARY STRUCTURE

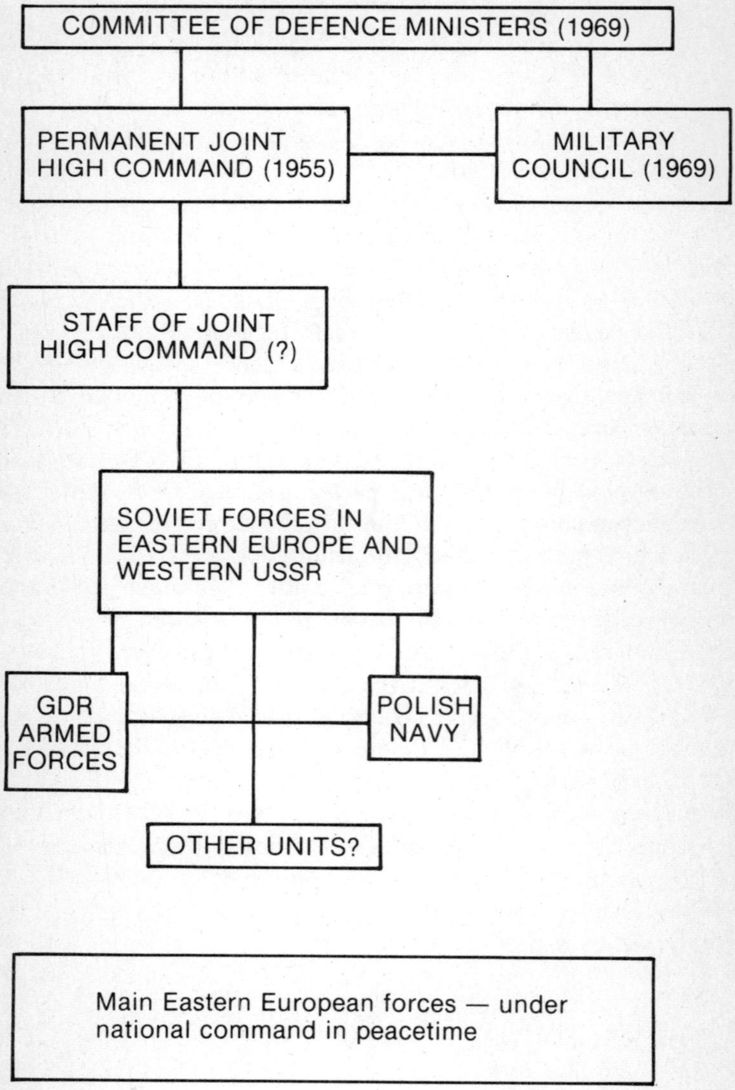

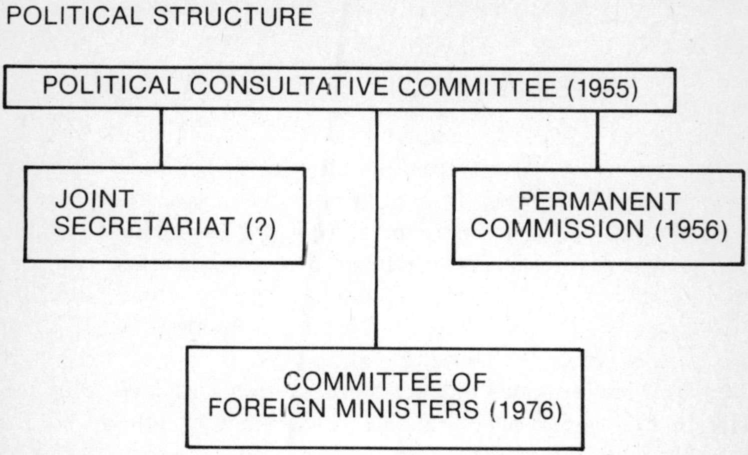

NB Relationships indicated by this table are tentative only–see text for further explanation.

FIGURE 7.1 *Principal institutions of the WTO (with dates of establishment where known)*

However, the distinction can be made as a way of organising an examination of the institutions.[9]

The Joint High Command (JHC) of the WTO was established by Article 5 of the original treaty, but the Committee of Defence Ministers (CDM) and Military Council were not set up until 1969. (The Political Consultative Committee or PCC, the senior political body, was established under Article 6 of the treaty and empowered to create additional bodies as the need arose; this is what happened at the March 1969 meeting in Budapest.) The WTO Commander-in-Chief, currently Marshal Viktor Kulikov, chairs meetings of the CDM, which includes his Chief of Staff, and the Ministers and Deputy Ministers of Defence of the WTO nations. The JHC and Joint Staff include senior officers from the member countries, while Soviet officers serve as representatives to military headquarters in the Eastern European states. The WTO Commander-in-Chief is one of three Soviet officers with the rank of First Deputy Minister of Defence, who are the immediate subordinates of the Soviet Minister of Defence himself. The Military Council, again chaired by the Commander-in-Chief, includes senior Soviet officers and Eastern European representatives. It appears that the Joint Staff prepares operational plans and manages field exercises, while the Military Council advises the CDM, and co-ordinates training and exercises. The CDM is the senior body, and supervises the Joint High Command. The WTO Commander-in-Chief and Chief of Staff have always been Soviet officers, although these WTO bodies have achieved rather more independence from Soviet structures than they once had. Originally part of the Soviet General Staff, the Joint High Command seems to have acquired its own staff and become an independent part of the Soviet Ministry of Defence at some time in the 1960s, while some analysts consider that the Joint Staff headquarters moved in 1972 from Moscow to Lvov in the Ukraine.

Despite their nominal independence, it seems unlikely that these WTO bodies would have wartime responsibilities to add to their peacetime tasks of dealing with exercises and training. When the multinational force occupied Czechoslovakia in 1968, it was commanded not by the WTO Commander, but by General Pavlovsky, the Commander of Soviet Ground Forces. In all likelihood, command of WTO forces would be exercised by the Soviet High Command if war broke out with NATO. In peacetime, however, the WTO Command controls Soviet forces in Eastern Europe and the Western USSR, as well as the entire German Democratic Republic armed forces, the Polish Navy (which reportedly forms part of a United Baltic Navy with GDR and Soviet Baltic fleets), and perhaps some other elite units like the Polish airborne division. Other Eastern European forces remain under national command in peacetime (a relaxation from the situation before 1955, when Soviet officers actually served in the Eastern European headquarters and ministries), while Romanian forces are not integrated into the system at all, and would apparently remain under Romanian command even in wartime.[10] The WTO air defence system has traditionally been controlled centrally from Moscow by the Commander of Soviet Air Defence Forces, although recent changes in Soviet defence organisation itself have complicated this picture.

The armed forces which make up the WTO's strength are heavily concentrated in the 'Northern Tier' countries (GDR, Czechoslovakia and Poland). These countries' forces are generally better trained and equipped than those of the remaining three 'Southern Tier' nations, and Soviet forces in Eastern Europe are concentrated in the northern sector. There are thirty one Soviet divisions permanently stationed in Eastern Europe – twenty in the GDR, two in Poland, five in Czechoslovakia and four in Hungary.[11] The stationing of these Soviet forces is governed by bilateral agreements between the Soviet Union and the respective governments. The Eastern European forces themselves have been affected by the crises which have plagued the WTO: the Hungarian and Czechoslovakian armed forces were sharply reduced after 1956 and 1968 respectively, and in Hungary remain smaller than they were before 1956.

In examining the way these structures have operated in decision-making on nuclear issues, it is instructive to look at the deployment of nuclear weapons with WTO forces. It has already been mentioned that the early 1960s saw a 'nuclearisation' of Soviet military thinking and forces. This was not a result of consultation between the Soviet Union and its allies, but a reflection of the position adopted by the Soviet leadership under Khruschev that nuclear weaponry would be the decisive factor in any future war. Nuclear-capable weapons systems were introduced into Soviet and Eastern European forces, including Su-7 aircraft (in Poland and Czechoslovakia), FROG and SCUD missiles (throughout the WTO).[12]

Since this is an area in which the Soviet Union is particularly secretive, it is impossible to say for sure whether the warheads for these weapons were kept in Eastern Europe or stored on Soviet territory. Western analysts have tended to assume that only Soviet forces actually trained to use nuclear warheads themselves, and that their warheads were not kept permanently in Eastern Europe, although some have considered that there is Eastern European involvement in nuclear preparations, at least by the GDR.[13]

Even if the Eastern European armies did not themselves have nuclear weapons, they seem to have been involved in training for a possible war in which Soviet nuclear weapons based in the Soviet Union itself might be used in Europe. The multilateral exercises which started in the 1960s seem to have been supervised by the Joint High Command, but it is not known whether the principles of the new strategy were discussed in, say, the Political Consultative Committee. Czechoslovakia and Romania, at least, felt that whatever discussion did take place was inadequate.

Since the 1960s, later generations of nuclear-capable weapons systems have been added to the inventories of WTO forces, although there was no public admission of the presence of warheads in Eastern Europe until late 1983. Among these new systems are aircraft like the MiG-27, Su-17/20 and Su-24, of which the Su-20 is in Polish service, and a range of nuclear-capable artillery, of which one model (the M-

1973 howitzer) is reported as in service with Hungary and the GDR. Once again, however, there is no reason to suppose that nuclear warheads are available to any forces besides those of the Soviet Union.[14]

The Eastern European states are, as far as we know, non-nuclear partners in a nuclear alliance in which the possession and possible use of nuclear weapons are Soviet prerogatives. Since military command is also largely in Soviet hands, it is important to look at the WTO's political structures to see if there is any further evidence of how decisions concerning nuclear weapons are taken.

POLITICAL STRUCTURES

The Political Consultative Committee (PCC), established by the terms of the Warsaw Treaty itself, remains the senior body of the WTO. Its communiques are signed by party leaders rather than military officers, and its membership includes, in addition: heads of government, foreign and defence ministers, the Soviet Chief of General Staff, and the Commander-in-Chief and Chief of Staff of the WTO itself. The PCC has an executive agency in the Joint Secretariat, with responsibility for armaments, logistics and preparation of agendas, and a Permanent Commission responsible for recommendations on foreign policy questions. The Committee of Foreign Ministers works with the other subordinate bodies to advise the PCC, and the presence of the WTO Commander-in-Chief on the PCC suggests that this is the forum in which he reports to political leaders on the work of the military bodies already described.[15]

A Soviet account of the PCC's work describes it as the consideration of 'questions of European security, ways of strengthening world peace, and the tasks of improving defence co-operation of the allied states'.[16] The Permanent Commission was set up by the PCC in January 1956, and the Committee of Foreign Ministers in 1976 (possibly as a consequence of the increasing complexity of foreign policy co-ordination in the era of detente). Some uncertainty surrounds the creation of the Secretariat – some sources date it from 1956, while others consider that it too was set up in 1976.[17]

PCC meetings alternate between the different WTO capital cities. The Treaty does not specify how often it should meet, and the regularity of its meetings has not been constant over the years. Its role in the conduct of foreign policy by the Soviet Union and its allies has been to co-ordinate policy and to make public statements embodying common positions. However, positions put forward by the PCC are not always simple reflections of Soviet policy. In 1958, for example, a PCC meeting endorsed a version of the Rapacki Plan for denuclearisation in Central Europe, which was originally a Polish initiative.

It is difficult to be sure about the methods used for arriving at decisions within the PCC. The fact that it did not meet at the height of the Hungarian or Czechoslovakian crises seems to indicate that in the 1950s and 1960s the Soviet Union preferred to use bilateral and *ad hoc* meetings to manage internal alliance problems, using WTO forums only for foreign policy issues where consensus was easier to reach. By 1981 the WTO seems to have been working more smoothly. Both the Committee of Foreign Ministers and Committee of Defence Ministers met in early December 1981, just before the declaration of martial law in Poland, and it looks as though the meetings were used to brief alliance members on the impending military takeover and to co-ordinate a united position.

The available evidence suggests that the creation of additional discussion forums after 1968 was designed to aid this process of easing internal tensions, and to make sure that Eastern European leaderships felt they were being adequately consulted on both military and political issues. The mechanisms were quite severely tested by the nuclear counter-deployments episode in 1983, and this is more fully examined in this chapter's concluding section. The benefits of this system for the Soviet Union are that public statements by the entire WTO help to legitimise Soviet foreign policy and sustain political cohesion, although the purposes for which political cohesion is required may vary over time. For example, in 1955 China was an observer at the signing of the Warsaw Treaty, but by 1968 the Sino-Soviet split had occurred and one of the WTO's functions was to make it clear that the Eastern European states were aligned with Soviet, not Chinese, policy.

Apart from its most publicly visible institutions, the WTO has a number of other institutional structures which illustrate once again the close ties between military and political aspects of the alliance. The co-ordination of military training involves emphasising the political signficance of several national armies training to fight side-by-side. It has been pointed out that the WTO's military strategy and training have the consequence of rendering the Eastern European armies unable individually to defend their own territory against Soviet intervention, which makes it possible for the Soviet Union to deter any political developments in Eastern Europe which might turn out to be contrary to its own interests. The WTO armies all have directorates with responsibility for political training within the forces, and all except Romania participate in the integrated officer education system whereby senior officers are trained in the Soviet Union itself.[18]

The practical significance of these structures for the senior military personnel who meet in the WTO's senior multilateral bodies is that they do not participate as products of differing national systems who must mediate between possibly diverging national interests. They would not have become members of these senior bodies without passing along politically similar paths which are designed to pre-empt divergences of interest. In this sense, the post-1968 reforms in the WTO have not put power into the hands of any potentially anti-Soviet elite, with the possible exception of Romania. Regular bilateral party, state and military contacts outside the WTO framework reinforce the multilateral structures, and one must assume that a good deal of groundwork is done before important meetings in order to prepare statements in advance. In these circumstances, it is perhaps surprising that evidence still emerges of dissension within WTO institutions. The next section of this chapter will examine one important recent example of the WTO's role in relation to Soviet nuclear weapons – the 1983–4 'counter-deployments' period.

COUNTER-DEPLOYMENTS IN EASTERN EUROPE – A CASE STUDY

The possibility of Soviet deployments of nuclear weapons in Eastern Europe to counter NATO weapons emerged as early as 1958, when Khrushchev raised the issue as a possible response to nuclear deployments with West German forces.[19] However, as has already been mentioned, no public confirmation of the presence of nuclear weapons in Eastern Europe occurred until late 1983, and Western sources have taken different views. The official US position has varied. At one time it was stated that there were none, and that consequently no value could be attached to WTO support for a Central European nuclear-free zone. Later it was argued that such weapons had been in Eastern Europe for some time, so that the WTO could not claim that the 1983 and 1984 counter-deployments were a new step.[20] Even if we do not know the truth about the pre-1983 situation, the WTO's handling of the counter-deployments issue makes an intriguing case-study.

Before NATO's own new medium-range weapons became a public

issue, there were rumours that in the mid-1970s some of the Eastern Europeans, notably Hungary and Romania, had expressed reservations about Soviet plans for the deployment of SS-20s in the Soviet Union.[21] Some Hungarian sources claim privately that Hungary actually opposed SS-20s at that stage. If this is true, it does not seem to have affected Soviet policy significantly. It is possible that the step taken by President Brezhnev in October 1979, a troop withdrawal from the GDR, was partly prompted by pressure from Eastern Europe to take an initiative which might head off likely Western moves towards stationing new weapons in Western Europe.

By the beginning of 1983, the WTO's response to the NATO 'dual-track' decision of December 1979 had taken shape in the form of a number of proposals made by Brezhnev and his successor, Yuri Andropov. However, it was the Soviet Union itself which made the negotiating proposals and also made the most explicit comments on likely Soviet counter-measures if NATO deployments went ahead. Statements from WTO bodies in 1983 tended to support the Soviet proposals without going further and threatening counter-measures.

In January 1983, the WTO Political Consultative Committee met in Prague and issued its 'Political Declaration of the States Parties to the Warsaw Treaty'. This expressed the WTO's 'appreciation' of proposals made by Andropov in December 1982, but did not say anything about possible counter-measures. A few days later, the GDR Politburo and Council of Ministers issued a statement which was more enthusiastic about Andropov's initiatives, which it 'greatly appreciated', but also mentioned the need to take 'whatever steps are required to safeguard. . . .defence capacity'.[22] The tone of the GDR statement suggests a strong loyalty to the Soviet Union, but also an awareness that the GDR had a particularly strong interest in seeing the Soviet proposals succeed, because of its own vulnerability if Soviet counter-measures were taken.

The Prague Declaration more or less set the tone for WTO statements during 1983. A further summit meeting took place in June 1983 in Moscow, but it is curious that this was not a formal meeting of any WTO body. It was described in communiques as 'a meeting of party and state leaders' from the WTO states.[23] The statement produced by this meeting stressed that the WTO would not permit NATO to achieve military superiority, but again made no specific mention of possible counter-measures.

'Having jointly analysed the situation developing in Europe and in the world as a whole, the participants in the meeting express on behalf of their socialist states concern over the continuing growth of tension, the further destabilisation of interstate relations and the growth of the threat of the nuclear war with its catastrophic consequences. They find it necessary to draw the attention of all countries and peoples to these dangers. . .Proceeding from the interests of peace and security, the states participating in the meeting declare that they will in no case allow military superiority to be achieved over them'.[24]

There was a good deal of speculation in the Western press over this meeting, since it was suspected that it had been called in order to produce a stronger statement than the one which actually emerged, but had been prevented from doing so by Romanian and Hungarian resistance.[25] The Soviet Union denied these Western reports, but the episode was a curious one, since the meeting seemed almost a return to the 1968 tactic of convening *ad hoc* meetings rather than formal WTO sessions to cope with a particularly urgent topic.

In October 1983, a meeting of the Committee of Foreign Ministers took place in Sofia, and the WTO again expressed approval of Soviet proposals, warned against NATO's precipitation of another round in the arms race, and insisted that NATO would not be allowed to gain superiority over the WTO.[26] However, Marshal Kulikov attended the meeting and was blunter in his language. According to an Associated Press report, he stated that if the NATO deployments went ahead, the Soviet Union would suspend its moratorium on medium-range weapons (announced by Brezhnev in 1982), deploy additional weapons after consultation with its allies, and strengthen WTO conventional forces.[27]

The pace of events in the second half of October was rapid – with an interview in the West German magazine 'Stern' with Colonel General Chervov of the Soviet General Staff, saying that Soviet tactical nuclear weapons *were* already in place in Eastern Europe, and would be modernised if NATO deployments went ahead; critical comments from President Ceausescu of Romania, implying that the Soviet Union would be at least partly to blame if the Geneva Intermediate Nuclear Forces talks broke down; statements from an 'extraordinary' session of the Committee of Defence Ministers and also from a COMECON session, the latter repeating the Soviet offer to destroy some Soviet missiles in return for the non-deployment of cruise and Pershing II.[28]

On 25 October, the Soviet Defence Ministry announced that preparatory work had been started for the deployment of new Soviet missiles in the GDR and Czechoslovakia, and this announcement was accompanied by statements by the Czech Communist Party Presidium, the Czechoslovakian government, the GDR Defence Council and a communique issued after a visit to Prague by the GDR Party Secretary Erich Honecker.[29]

Similarly, when President Andropov announced the Soviet decision to break off the Geneva negotiations and to take a number of countermeasures, including speeding up work on the Czechoslovakian and GDR preparations (on 24 November), he made the announcement in his capacities as General Secretary of the Soviet Communist Party and Soviet President.[30] The Eastern European counter-deployment preparations were said to be in progress 'in agreement with the governments of the GDR and Czechoslovakia', rather than by agreement among the WTO governments. It is interesting that Andropov's statement itself does not specifically identify the weapons intended for the GDR and Czechoslovakia as nuclear, though the implication is that this is what they are.

This sequence of events during 1983 illustrates that once it became clear that NATO's deployments would go ahead in Western Europe, the WTO's collective diplomatic efforts gave way to bilateral military arrangements between the Soviet Union, the GDR and Czechoslovakia. (It is also noticeable that the last straw, in the Soviet Union's view, seems to have been the West German Bundestag vote on 22 November to reaffirm support for NATO deployments, rather than the arrival of cruise missiles at Greenham Common in Britain, which occurred on 14 November.) Supportive statements made subsequently in the GDR and Czechoslovakia did speak of the measures as taken to ensure the protection of the entire WTO, but they confirmed that the agreements were bilateral.[31]

This resort to bilateralism in military relations within the WTO might have been expected. It has already been noted that the presence of Soviet forces in Eastern Europe depends on bilateral treaties rather than on the text of the Warsaw Treaty itself, although there are presumably also unpublished annexes to the treaty which may cover this issue. We do not have evidence on the precise wording of the 1983 agreements with the GDR and Czechoslovakia, but there is no reason to suppose that the counder-deployments are, any more than the presence of Soviet troops, legally or technically dependent on the *WTO* at all.[32] Even so, it was obviously important that WTO bodies should endorse them once the announcement had been made by Andropov.

A meeting of the Committee of Defence Ministers in Sofia in December 1983 did endorse the measures, although reports of Romanian unease persisted.[33] A meeting of WTO foreign ministers in Stockholm on 16 January 1984 did not mention the issues, although it was not a formal WTO meeting, and took place in connection with the CSCE Conference.[34] When the Committee of Foreign Ministers did endorse the counter-deployments in April 1984, it used only lukewarm language in referring to the counter deployments and noting the breaking-off of the Geneva talks: 'This has compelled the Soviet Union to adopt a number of response measures. Talks on nuclear armaments have been terminated.'[35] This conspicuous absence of enthusiasm was a sharp contrast with language being used in the Soviet press in early 1984; a January article in the army newspaper Krasnaya Zvezda spoke of the 'unshakeable resolve' of Soviet missile-men stationed in the GDR.[36] No Political Consultative Committee meeting took place between 1983 and Mikhail Gorbachev's announcement of a freeze on Soviet counter-deployments in April 1985, so they had not in fact been endorsed by the WTO's senior body at that point. However, the illness of Konstantin Chernenko probably caused plans for such a meeting to be put off in late 1984.

The evidence of the 'counter-deployments episode' suggests that the Eastern European states in the WTO were concerned during 1983 and 1984 to limit the damage to East-West relations caused by the failure of US–Soviet talks in Geneva. They were less prepared than the Soviet Union to stress the military counter-measures which were seen as being necessary, and even after the counter-deployments became a fact of life, indications of concern about them emerged. The military arrangements made for the counter-deployments themselves were bilateral.

Even loyal allies of the Soviet Union like Erich Honecker of the GDR and Lubomir Strougal of Czechoslovakia expressed some degree of disquiet at the end of 1983, and this was occasioned by a mixture of military, political and economic concern. In military terms, the counter-deployments made little sense, since they made Eastern Europe more vulnerable to attack, and strained the credibility of the Soviet Union's stated position on No First Use of nuclear weapons.[37] In political terms, the whole deployment and counter-deployment episode damaged East–West relations. In economic terms, it appears that the counter-deployments imposed a burden on the whole of the WTO. Comments made by Heinz Hoffmann (GDR Defence Minister) and Todor Zhivkov (Bulgarian party leader) in late 1983 suggested that their cost would be spread throughout the WTO.[38]

During the course of 1984 further statements were made by the Soviet Union on the continuation of the counter-deployment programme, and Western analysts debated what combination of SS-21s, SS-22s and SS-23s it consisted of. By April 1985, when the Soviet moratorium was announced, the US Government and NATO were in agreement that SS-21s and 22s had been deployed, but not SS-23s. Figures given in early 1985 for these Eastern European counter-deployments were less specific than for SS-20s in the Soviet Union itself, but claimed 60 SS-22s with no precise estimate for SS-21s.[39] With such uncertain figures it is perhaps risky to speculate, but the rate of deployment seems slow by comparison with that for cruise and Pershing II in Western Europe. It is not unreasonable to suppose that if there were more counter-deployments in Eastern Europe than there are cruise and Pershing II in the West, much would have been made of the fact. If there are not, this may be evidence of Eastern European reluctance to see the counter-deployment programme proceed too quickly, despite the public position that 'parity' will be maintained.

The 'counter-deployments episode' provides the only recent occasion on which Western observers have been able to follow the public evidence of a major decision concerning nuclear weapons deployment being taken within the Warsaw Treaty Organisation. It is reasonable to ask how typical of WTO decision-making procedures the episode was.

The main distinguishing feature of the episode was the degree of publicity involved. Before late 1983, the Soviet Union had simply never confirmed the presence of nuclear weapons on Eastern European territory. The high public visibility of the announcements made in 1983 and 1984 was an inevitable consequence of the central role in East-West relations assumed by land-based theatre nuclear weapons during the 1979-83 period. The Soviet Union had repeatedly said that it would respond to cruise and Pershing II, and so was committed to doing so in a highly public way.

The public posture, however, did not seem to make agreement within the WTO any easier. The episode revived worries about nuclear consultation within the WTO which had arisen twenty years previously. It also gave rise to concern in the minds of Eastern European leaders who had traditionally been thoroughly loyal to the Soviet Union. The worries voiced by Honecker and Strougal reflected their awareness that their own populations had over the previous few years been realerted to the dangers of nuclear weapons by the WTO's own publicity directed against NATO programmes.

The role of the Eastern European states within the WTO during this period seems to have consisted of attempts to moderate the Soviet response and to facilitate an agreement at Geneva. When the talks broke down, the Soviet Union turned to bilateral military mechanisms between itself and the GDR and Czechoslovakia. This combination of collective diplomacy and bilateral measures bears some similarities to NATO's management of US–allied nuclear relations, with its use of Programmes of Co-operation (see Chapter 6 on NATO), though the WTO's arrangements do not seem to allocate nuclear warheads to the Eastern Europeans themselves.

This is not to say that the counter-deployments were entirely reactive to NATO. The new Soviet missiles are certainly programmes which would have been in progress irrespective of NATO plans, and they represent 'modernisations' of weapons systems some of which may have been in Eastern Europe, possibly with nuclear warheads available, before 1983.[40] The attendant publicity, rather than the military arrangements or the principle involved, was the new element.

It was never likely that the WTO's decision-making procedures would have reached the critical point at which the GDR or Czechoslovakia would have refused to accept Soviet counter-deployments on their territory. The established mechanisms described earlier in this chapter have been designed to ensure that no such crisis point is reached. However, it is possible that the public nature of the issue gave the Eastern European leaderships some leverage over the Soviet Union even if they could not directly challenge Soviet policy. Even within WTO institutions, the defence ministers seem to have endorsed the counter-deployments with more alacrity than the foreign ministers, who must have been all too aware of the frozen state of East-West relations throughout most of 1984.

It is possible that the Eastern European leaderships have exercised some leverage on the Soviet Union over the return to US-Soviet negotiations and the renewal of the Warsaw Treaty itself. In early 1985, reports filtered into the Western press that negotiations over the renewal of the treaty, due in May, were proving difficult. Mikhail Gorbachev announced the Soviet moratorium on SS-20 and Eastern European deployments on 8 April, and most attention was focused on the part of it which affected SS-20s. It is true that a freeze on SS-20s at such high levels is virtually meaningless, but a freeze on Eastern European counter-deployments might be more significant, since it represents some restraint on a more recent programme which would otherwise, presumably, have grown considerably in size.

Although one must assume that withdrawals from the WTO were

never seriously contemplated, it may not be too far-fetched to suggest that the less widely-publicised part of the Gorbachev initiative was something which some of the Eastern European leaderships might have pressed for as a *quid pro quo* for the renewal of the treaty. Doubtless Gorbachev would have taken an initiative in any case, with an eye on Western European opinion and with the aim of putting NATO on the defensive, but the Eastern European leaderships may well have been eager to see an initiative taken which would ease their own positions and might repair the overall state of East–West relations. In April 1985 the Warsaw Treaty was renewed for twenty years, with an option to renew it for a further ten years after that, to the year 2015.[41]

NOTES

1. J. F. Brown 'The Future of Political Relations Within the Warsaw Pact', in *The Warsaw Pact: Alliance in Transition?* David Holloway and Jane M. O. Sharp (eds), (Macmillan 1984) hereafter referred to as *Holloway and Sharp*.
2. The term 'Eastern Europe' is sometimes contentious; it us used here simply to refer to the six states which are the Soviet Union's allies in the Warsaw Treaty Organisation – Bulgaria, Czechoslovakia, GDR, Hungary, Poland and Romania.
3. Terms such as 'military doctrine' and 'military strategy' have specific meanings in Soviet military contexts, which do not necessarily correspond to Western usages. For further explanations, see ch. 3 of D. Holloway *The Soviet Union and the Arms Race* (Yale University Press, 1983) and ch. 3 of H. F. and W. F. Scott *The Armed Forces of the USSR* (Westview Press, 1979).
4. The text can be found in M. Mackintosh 'The Evolution of the Warsaw Pact' *Adelphi Paper*, no. 58 (June 1969).
5. For further examination of these motivations see: *Adelphi Paper*, no. 58 (op. cit., footnote 4); M. Mackintosh, 'The Warsaw Treaty Organisation: a History' in Holloway and Sharp; R. A. Remington, *The Warsaw Pact: Case Studies in Communist Conflict Resolution* (MIT Press, 1971); F. Fejto, *A History of the People's Democracies: Eastern Europe Since Stalin* (Penguin: 1977).
6. For example – V. Alexandrov, *The Warsaw Treaty and Peace in Europe*, (Novosti Press, 1980).
7. These points are dealt with in: T. W. Wolfe, *Soviet Power and Europe 1945–70* (Johns Hopkins Press: 1970) and S. M. Meyer, 'Soviet Theatre Nuclear Forces' *Adelphi Papers*, nos 187 and 188 (1983–84).
   On the Eastern European reaction: J. Sharp, 'Security Through Detente and Arms Control' in Holloway and Sharp; D. Burke, 'Defense and Mass Mobilisation in Romania', *Armed Forces and Society*, vol. 7, no. 1 (Fall 1980); and C. Rice, *The Soviet Union and the Czechoslovak army 1948–1983: uncertain allegiance* (Princeton University Press, 1984).

8. The opposition to increases was largely Romanian, although there may have been support for the Romanian position from other quarters. See contemporary reports in: *The Economist* (2 Dec. 1978); *Financial Times* (28 Nov. 1978 and 15 Aug. 1979).
9. Sources of information for this table and for the account which follows include: *The Military Balance* (IISS London); D. Isby, *Weapons and Tactics of the Soviet Army* (Jane's: 1981); F. Wiener, *The Armies of the Warsaw Pact Nations* (Carl Ueberreuter Publishers: 1981) and M. Mackintosh, 'The Warsaw Treaty Organisation: A History' in Holloway and Sharp.
10. See ch. 4 of I. Volgyes, *The Political Reliability of the Warsaw Pact Armies: The Southern Tier* (Duke Press Policy Studies: 1982).
11. In addition to the works cited in footnote 9, further information can be found in: W. J. Lewis, *The Warsaw Pact: Arms, Doctrine and Strategy* (McGraw-Hill Productions: 1982) and R. W. Clawson and L. S. Kaplan (eds) *The Warsaw Pact: Political Purpose and Military Means* (Scholarly Resources Inc., 1982).
12. *The Military Balance 1984–5* (IISS: 1984) pp.135–6.
13. Bill Arkin considers that the GDR army has nuclear weapons storage sites. See his 'Nuclear Weapons in Europe', in *Disarming Europe,* M. Kaldor and D. Smith (eds) (Merlin: 1982) pp.35–63.
14. Data from: *The Military Balance 1984–5; Soviet Military Power 1985* (US Department of Defence); P. Rogers, *Guide to Nuclear Weapons 1984–5* (University of Bradford School of Peace Studies, 1984).
15. The essentials of this account are taken from *The Military Balance 1982–3.*
16. V. Alexandrov, op. cit. in footnote 6, p.21.
17. M. Mackintosh and T. O. Cason, ('The Warsaw Pact Today: the East European Military Forces' in Clawson and Kaplan) favour the earlier date; J. Dean ('The Warsaw Pact in the International System' in Holloway and Sharp) and Valentin Alexandrov give the later one.
18. These structures are thoroughly examined in: C. Jones, *Soviet Influence in Eastern Europe: Political Autonomy and the Warsaw Pact* (Praeger: 1981).
19. See T. W. Wolfe, *Soviet Power and Europe 1945–70,* p.151.
20. Pointed out by J. McMahan, *Reagan and the World: Imperial Policy in the New Cold War* (Pluto: 1984) p.65.
21. Reported by J. Sharp, 'Security Through Detente and Arms Control', in Holloway and Sharp.

    The SS-20 is a mobile, triple-warhead intermediate-range nuclear missile, and was first deployed in 1977. It represented a 'modernisation' of SS-4 and SS-5 missiles, deployed from 1959 and 1961 respectively.
22. The Prague Declaration text was published in English by Panorama DDR, with the GDR statement in the same pamphlet.
23. See text in *International Affairs* (Moscow, Sept. 1983).
24. *International Affairs* (Moscow, Sept. 1983).
25. See contemporary reports in: 'Romania Thwarts Soviet Plan to Counter Cruise', *Guardian* (29 June 1983); 'Soviet Leaders Fear NATO Will Take Advantage of Warsaw Pact Discord', *The Times,* (30 June 1983); 'Soviet Denies Reports of a Split in Warsaw Pact Over Arms Policy', *International Herald Tribune* (30 June 1983).

26. Text in *Soviet News* (London, 26 Oct. 1983).
27. Report in *Guardian* (14 Oct. 1983).
28. Press reports of Chervov's interview and the Romanian statements: GND/DPA/UPI/REUTER report (Bonn, 17 Oct. 1983); *Guardian* (19 Oct. 1983); *Financial Times* (26 Oct. 1983); CDM and COMECON texts both in *Soviet News* (26 Oct. 1983).
29. Reports and text in: 'US Foreign Broadcasts Information Service (FBIS) Report' (25 Oct. 1983); BBC Monitoring Service Summary of World Broadcasts (26 Oct. 1983).
30. *Soviet News* (30 Nov. 1983).
31. For example: 'Rude Pravo' editorial (12 Dec. 1983) in *FBIS Report* (14 Dec. 1983) and speech by GDR Prime Minister Willi Stoph on 8 Dec. 1983 in *FBIS Report* (9 Dec. 1983).

    A press conference was also given in Moscow on 5 December by Marshal Ogarkov, then Soviet Chief of General Staff *(Soviet News,* 7 Dec. 1983).
32. The troop-stationing agreements, and other bilateral treaties between the WTO states, are summarised in *Keesing's Treaties and Alliances of the World* (Longmans, 1983) section 11.
33. *Guardian* (8 Dec. 1983); *International Herald Tribune* (22 Dec. 1983).
34. *Soviet News* (18 Jan. 1984).
35. *Soviet News* (25 Apr. 1984).
36. *Soviet News* (18 Jan. 1984).
37. As pointed out in: *No First Use,* J. Goldblat, S. Lodgaard and F. Blackaby (eds) (SIPRI, 1984) p.20.
38. See R. English, 'Eastern Europe's Doves', *Foreign Policy* (Fall 1984) pp.44–60.
39. These western assessments are in: 'NATO Nuclear Planning Group Final Communique' (NATO Press Service, 12 Oct. 1984) and *Soviet Military Power* (1985).

    Gorbachev's statement in the form of an interview with a Pravda correspondent, is in *Pravda* (8 Apr. 1985). English text in *Soviet News* (10 Apr. 1985).
40. For details on the weapons and their predecessors see: *Jane's Weapons Systems 1984–5,* (Jane's, London) and J. M. O. Sharp, 'Soviet Response to Cruise and Pershing', *Bulletin of the Atomic Scientists* (Mar. 1984) pp.3–4.
41. *Soviet News* (1 May 1985).

Postscript

The task of this book has been to map out, as accurately as possible given the amount of published information available, how decisions on nuclear weapons are made. The structures and processes of defence decision-making in each of the nuclear weapons nations and two major alliances have been examined separately. Throughout these chapters, however, a number of similarities have emerged, themes which have come up sufficiently often that they cannot be ignored.

The first is the importance of the early stages in the life of a weapons system. The key decisions are made long before a prototype has been built as in France, for example, where the decision to begin applied R & D is tantamount to the decision to produce and deploy the weapon. The funds, and more importantly, the personnel committed to a project in the design stages, are greater than is generally realised. This momentum builds up, involving the commitment of laboratory scientists, designers, contractors and military personnel, before the formal decision-making procedures come into operation. These scientists also appear to hold effective lobbying power, as in the United States where the heads of the three weapons laboratories are credited with turning the tide in Washington against a Comprehensive Test Ban Treaty.

Rivalry between the services is another key factor in pushing forward weapons decisions. Now that nuclear weapons are designed in different forms–as artillery shells, depth charges and tactical missiles as well as intercontinental missiles and air drop bombs–the navies and air forces of these nations (and in some cases the armies and marine corps as well), have competed to own the newest technology as part of their capabilities. American sources are very much more forthcoming about this than other nations, where the rivalry is more apparent between the lines. Competition is inherent in military thinking, and efforts to combine military advice-giving have consistently failed. Speaking of the advice given by the US Joint Chiefs of Staff, former Defence Secretary Harold Brown told Congress in 1982 that it was 'almost without exception not very useful or the reverse of being helpful. That is, worse than nothing'. This means that civilian decision-makers have been receiving unco-ordinated competing military advice from their separate services (this is particularly notable in China), oriented far more to individual weapons systems than to strategy and policy as a whole.

Thirdly, executive financial control of weapons development appears, in those countries where information is available, to be weak. The structures for interdepartmental budgetary evaluation of nuclear systems are understaffed, and ill-equipped for the job in terms of technological expertise. Defence departments can in practice move funds between programmes during the financial year without requiring the approval of treasury officers. In most nations, defence departments are not obliged to disclose to elected representatives what they plan to spend on specific new weapons systems, with the result that by the time the public is aware of a weapons system, so much has already been spent that the project is virtually unstoppable.

Foreign policy formulation today is more dominated by weapons developments than in the past. This becomes particularly clear in the studies of NATO and the Warsaw Treaty Organisation, and in Britain where, for example, the Foreign Office must clear arms control policy with the Ministry of Defence, but the Ministry of Defence is not obliged to clear new weapons developments with the Foreign Office. Arms control does not operate at the crucial early stages of weapons development. Both super-powers having developed new systems have used their deployment as 'bargaining chips' in negotiations supposed to reduce numbers of that same system. Arms control departments have suffered budget and other deprivations in recent years, while weapons budgets have increased massively. Funding for arms control research is negligible compared to military research funding. The 'foreign threat', based on the interpretation of intelligence data by defence and foreign service intelligence agencies, is a more potent influence on new weapons starts than is arms control.

Another recurring theme throughout these studies is that of secrecy. Although this seems a rather obvious point, and although a certain restriction of information on defence issues is clearly necessary, the effects on decision-making of the present extent of secrecy must be considered. As things stand, those who may have genuine and useful contributions to make on defence and nuclear decisions are not consulted, sometimes with disastrous results.

In the Soviet Union, because of secrecy, early and very crucial weapons developments are undertaken by scientists and designers without the scrutiny and criticism of their work by their peers which would help to obviate bad decisions; and exactly the same is true in the West. Throughout the development of a nuclear system, information is confined to a very few key personnel, who tend to operate in a closed world. In a striking number of instances, extending to think tanks and strategic planning advisers, these individuals regard themselves as the holders of information which citizens and politicians are incapable of understanding and they indeed speak in a language more comprehensible to their counterparts in other nations than to their fellow countrymen and women.

The closing of this world, focusing on ever-narrower technological complexities, results in broader policy options, such as non-nuclear and alternative defence strategies, not being considered.

'The scientists who work in the defence departments of governments, or defence industries, are not apostles of peace', wrote Lord Zuckerman, who was for a decade Chief Scientific Adviser to the British Ministry of Defence. 'Political and military leaders should cease seeking shelter behind the backs of those "experts" who take what is usually called the harder line. In the twenty years since the first major effort was made to bring the nuclear arms race to an end, masses of water has flowed under the bridge. If the bridge itself is not to become submerged, the politicians will have to take charge of the technical men.'

The sixth theme which recurs in these chapters is the power of the permanent bureaucracies. The development lifespan of a nuclear delivery system and its warheads is now fifteen or twenty years; because of the extreme sophistication of its components, dogged co-ordination is necessary through the feasibility studies, evaluation, prototype building, testing and production stages. These stages are in the hands of permanent bureaucracies, while government ministers, in the Western nations, are usually in office for periods of no more than two or three years; the pressures on them to maintain consistency with their predecessors' decisions are very great. Since the technology is so complex, the formal decisions taken in Cabinet or Politburo depend to a large extent on the options presented by their permanent secretariats. The parameters of these options do not normally include consideration of alternative defence policies. Fundamental questioning of the assumptions underlying particular decisions do not normally form part of the debate.

Any new administration or government coming to power has to take account of the realities of the *status quo*. These realities may concern funds already committed, unofficial agreements entered into, the pressure of alliances, the persistence of officials, or simply the lack of procedures to do what they want to do – but they are so real that they have left election manifestoes in tatters. This outcome of the studies, the realisation than an elected government may find it extremely difficult in practice to reverse nuclear decisions, has been one of the bigger surprises.

Many members of the public are under the impression that their elected representatives can affect nuclear policy. The final point to be made is that this is not so. Parliaments and elected representative assemblies in East or West have very little influence on nuclear weapons issues. In the United States, which has the greatest freedom of information and the most rigorous system of congressional control of funding allocations, the reality is that Congress barely alters what comes from the Pentagon in military nuclear requirements.

'The military Services can find ways to circumvent even the explicity stated intention of Congress', wrote Ted Greenwood. 'Amounts of money can be re-programmed; definitions and descriptions can be changed so that proscribed activities can be carried out under different names and different mission requirements. The time frame of government bureaucracies is rather long. They expect to be operating long after the attention of legislators or administration officers, who may be temporarily interested in their activities, has moved onto other things.'

The similarities between the five nations discussed in this book are striking in many ways. Most particularly so in terms of how little the public of each country knows about the key decisions taken when they are taken, or even afterwards, and how little its representatives, elected democratically or otherwise, are able to influence nuclear policy. In fact, the lack of effective review process in the West, and the absence of public accountability of those who do shape nuclear decisions, offer more similarities than differences with the process of nuclear weapons decision-making in the Soviet Union and China.

The aim of this book has been to provide an initial guide to how decisions on nuclear weapons are made. A second book will take up the themes mentioned above and examine them more closely, looking further into the unofficial methods of decision-making which underly the official ones. This book, by charting the official structures and processes, may assist those outside the systems to understand how they work, how similar they are, and how much they are interlinked. It may go some way to explaining why massive civilian demonstrations against new nuclear systems have had very little effect. It may enable those who work within the system to discern links or comparisons which had not previously been apparent to them. It may assist those in politics to do what Eisenhower and Macmillan urged them to do, to take hold of runaway technologies, scientific passions, military rivalries, corporate profiteering, bureaucratic persistency and international pressurising and subject them to wise policy-making.

Index